The Story of Canadian Roads

University of Toronto Press

Edwin C. Guillet

The Story of Canadian Roads

© University of Toronto Press 1966
Toronto and Buffalo
ISBN 0-8020-1414-3
LC 66-31427
Reprinted 1968, 1973
Printed in Canada

Generous assistance towards the costs of preparation and research for this history
of Canadian roads was provided by the Canadian Good Roads Association
during the Jubilee Year of the Association, 1964, in order to enable publication
by Canada's Centennial year, 1967.

Preface

THE PLAINEST and most far-reaching causes of social change are sometimes overlooked. While historians examine political upheavals and technological revolutions for the determinants of a society, they may pass over evidence of the slow spread of innovation, the gradual displacement of old customs by new, which profoundly affect the thought and behaviour of a people. Every nation has undergone such alterations, and comparison of the present with the past of one or two hundred years ago reveals their unexpected extent and influence.

Just this sort of social revolution, scarcely heralded by the writers of history, may be traced in the story of Canadian roads. Good highways are common today, and those communities which have seen their furthest development take them most for granted. Yet in their history can be read the history of a nation, a history shaped in part by the problems Canada has faced in seeking to connect two oceans by road. Those problems would have been formidable even in a land more densely populated: vast distances through forests and over plains; extended mountain ranges; rocks in some areas, permafrost in others; muskeg, snow, and ice; and, most recently, the straits and tides of the Atlantic Ocean.

Something of the considerable effect which these obstacles and the overcoming of them had on the spirit of the country can be seen by observing the changing conceptions of mobility over the relatively brief period of recorded history in Canada. The Indians subsisted without the wheel: they travelled by water and on foot, and in the east, when loads had to be transported overland, women were generally the beasts of burden. It was by the labours of women that the great Indian guide Matonabbee was able to lead Samuel Hearne across the Barren Lands to the mouth of the Coppermine River in 1771. "Women were made for labour," said Matonabbee. "One of them can carry or haul as much as two men can do. . . . There is no such thing as travelling any considerable distance, or for any length of time, in this country without their assistance."

Although the European settlers brought the wheel with them, they too avoided overland routes whenever possible, for trails were difficult to clear and almost as difficult to travel. They preferred to go in canoes and flat-bottomed boats, or by sleigh in winter over the frozen waterways. The first great improvement in transportation was the building of canals. Gradually and laboriously the most necessary roads were blazed, linking each new frontier with the areas already settled, helping to open lumbering districts, facilitating the delivery of mail, carrying manufactured goods to the backwoods and farm produce to the city. The internal combustion engine brought a new revolution. Freed at last from the restrictions imposed upon him by his dependence on muscle power, man achieved unprecedented mobility. With the motor car, for the first time, the ordinary man could travel whenever he wished, at his own speed, on the route of his choice across the great areas of a country that sprawled over half a continent. He could do all this provided he had roads adequate to the needs of the automobile. Thus roads were developed still further, and now the first efforts were made to bring national unity to the various local highway systems. In 1914 the Canadian Good Roads Association was formed—a body which since that time has had a great and salutary influence on the development of roads in this country.

In spite of such progress, the primitive often survived along with the modern. Even on main highways in the east in the early decades of this century, bridges over rivers and creeks were frequently washed out or dangerous to use, and fords and ferries were often preferable, if not essential. On the prairies, bridges across rivers were the exception rather than the rule. But the change of outlook brought about by the improvements already achieved had its effect on the future; good roads had become a necessity. Even in the most difficult areas they are now feasible instead of visionary, thanks to improved machinery and construction methods.

It is a sign of the times that the efforts and revenues of this nation are devoted so largely to the roads that unite it. Many years have passed since roads were considered purely a municipal responsibility. They have long been recognized as one of the prime concerns of every provincial government, and they have increasingly concerned the federal government as well. As the need for wider and better land communication has grown, so too have the extent and influence of roads. Today no other country, in relation to its size and population, is better served.

Materials for this book have been obtained on both sides of the Atlantic and all across Canada, and the acknowledgements are accordingly far-ranging. Some fifty travel books—many of them rare or unique—describing early conditions and experiences in western Canada were located in the Library of the British Museum, London. Other documents, photographs, and information were provided with the generous assistance of the Public Archives, Ottawa, and of librarians and archivists in all provinces. Provincial ministers of highways and of public works, aided by their staffs, have provided documents, typescripts, reports, and other publications which, perhaps more than any other source, have made a comprehensive story of Canadian roads possible.

Special thanks are due the Canadian Good Roads Association for sponsoring this book and for providing a great deal of important technical information. The CGRA's managing director, C. W. Gilchrist, has offered much encouragement, and has kindly furnished me with reports, bulletins, photographs, and other published and unpublished material of great value.

I am particularly indebted to Ian Montagnes for his capable help in guiding the manuscript through the press. He has contributed research and illustrations as well as editorial assistance, and the book has profited by his interest in it.

Many others have provided valuable assistance in various forms. David Pickard spent considerable productive time doing research on the modern period. Across the nation public officials and the staffs of museums, galleries, archives, and libraries provided official documents, typescripts, photocopies, illustrations, and rare books which enrich this history. As my wife and I visited each province to seek out the heritage of the past we were welcomed with courtesy and generosity. Those who personally provided material of special value are acknowledged here:

OTTAWA
W. Kaye Lamb, Dominion Archivist; Pierre Brunet, Assistant Archivist.

ALBERTA
E. J. Holmgren, Provincial Archivist; A. C. McClelland, Department of Highways, Edmonton; T. R. McCloy, Librarian of the Glenbow Foundation; Miss S. Jameson and Fred B. Johns of the picture collection of the Glenbow Museum, Calgary.

BRITISH COLUMBIA
Willard Ireland, Provincial Archivist; Miss Barbara McLennan, in charge of the picture collection of the Provincial Archives, Victoria; R. J. Baines, Administrative Assistant to the Minister, Department of Highways, Victoria; Mrs. J. Gibbs, City Archives, Vancouver.

MANITOBA
George Collins, Deputy Minister of Public Works; L. W. Blackman, Assistant Deputy Minister of Public Works; A. D. Maltman and Patrick Hunt, Administrative Officers of the Department of Public Works, Winnipeg; Hartwell Bowsfield, Provincial Archivist.

NEW BRUNSWICK
W. T. Hargreaves, Deputy Minister of Public Works; R. H. Sweet, Engineering Administration Co-ordinator, Department of Public Works, Fredericton; Mrs. M. Robertson, Department of Archives in the New Brunswick Museum, Saint John.

NEWFOUNDLAND
Hon. F. W. Rowe, Minister of Finance and President (1964) of the Canadian Good Roads Association; C. A. Knight, Deputy Minister of Highways; M. P. Murphy and N. C. Crewe of the Newfoundland Archives, St. John's.

NOVA SCOTIA
J. L. Wickwire, Deputy Minister of Highways; C. Bruce Fergusson, Provincial Archivist; Donald K. Crowdes, Director of the Citadel Museum, Halifax.

ONTARIO
J. D. Millar, Deputy Minister of Public Works; H. S. Howden, Assistant Deputy Minister of Highways; W. J. Fulton, former Deputy Minister of Highways; N. D. Bennett, Centennial Project Consultant, Department of Highways, Downsview; A. A. Walters, Historical Research Officer, Department of Highways; D. F. McOuat, Provincial Archivist; James L. Baillie, Royal Ontario Museum, Toronto.

PRINCE EDWARD ISLAND
R. Gordon White, Deputy Minister of Highways; Moncrieff Williamson, Director of the Confederation Art Gallery and Museum, Charlottetown.

QUEBEC
Hon. Gérard Morisset, Minister of Cultural Affairs, Quebec Museum; Roger J. LaBrèque, Deputy Minister of Roads; Joseph Matte, Assistant Deputy Minister of Roads.

SASKATCHEWAN
L. T. Holmes, Deputy Minister of Highways and Transportation; George Parker, Public Relations Officer, Department of Highways and Transportation, Regina; Allan Turner, Provincial Archivist; E. C. Morgan, Provincial Archives, Regina.

Contents

To Mary Elizabeth

my wife and travelling companion

The Story of Canadian Roads

The canoe was an essential part of Indian life in eastern Canada. In a culture without the wheel, the light bark craft was the only convenient means of transportation. The Micmac Indians of the Atlantic provinces, as shown in this painting by an unknown nineteenth-century artist, used it for hunting and fishing as well as for travel.

1

Canoe Routes and Portage Roads

THE FIRST ROADS IN CANADA were Indian trails connecting rivers and lakes, or by-passing rapids and waterfalls too dangerous for boats to travel. The early explorers and fur-traders called them roads, or "Indian paths," but in truth they were rarely more than openings in the woods just wide enough for a man to get by with a pack on his back or a canoe on his shoulders. They began as blazed trails, and because no one took time to improve them they remained simple footways that met the barest needs. Trees and rocks were skirted rather than removed, though the underbrush probably was cleared, just as today's hikers break off branches which lie in their way. Yet from frequent use the trails were often worn several inches, sometimes even a foot, into the light soil.

One of the most famous of these carrying-places, or portages, joined Lake Simcoe and Lake Ontario over the Humber-Holland trail. Although this portage was nearly thirty miles long, the Indians used to cross it in a single day, even when carrying heavy loads. We have been left a vivid picture of their speed and strength from Alexander Henry, a fur-trader who was taken over the portage in 1764 as an unhappy captive. His party had already paddled twenty miles across Lake Simcoe before entering the thick forest. It was a hot June day, and the woods and marshes along the trail swarmed with mosquitoes. Henry was carrying a pack he estimated weighed over a hundred pounds. The Indians were similarly burdened, but they walked so quickly that by ten o'clock next morning they had reached the shores of Lake Ontario, not far from the present site of Toronto. "It was a hard march," Henry wrote afterwards, "but I could by no means see myself left behind." He did not explain whether his reluctance rose from a sense of racial rivalry or a fear of being abandoned in the woods.

At Lake Ontario Henry's captors set to work immediately building a set of elm-bark canoes, for they had left their original craft at the other end of the portage. It was easier to make boats than to carry them thirty miles overland. Indeed, land travel at that time was invariably arduous. In early Canada man could move comfortably only on the water—and fortunately the country abounded in waterways. The lakes and rivers of the Canadian Shield and the southern lowlands formed a giant network feeding into the St. Lawrence. Along these silver roads the tribesmen, the explorers, and the fur-traders penetrated an otherwise inaccessible wilderness of rock and forest.

To this end the Indian had developed a superb craft. Canoes took many forms. Some were crude dugouts, hollowed from a single log. In the far west this method developed into an art: the huge coast cedars were made into dugouts up to sixty feet in length, decorated with carving and high figureheads, capable of carrying warriors and fishermen hundreds of miles along the Pacific shore. In the east the Iroquois built canoes of elm bark which could be stripped from the trees in thick sheets. This form of canoe was heavy, however, and might wear out in two or three years. By far the luckiest tribes were those who lived where the birch tree flourished. Its bark was the perfect canoe material, light, easy to work, and durable. Birchbark canoes, unless severely damaged, could last for twenty years.

Hardly more than a knife was needed to build one of them. The best

The birch-bark canoe was easily damaged, but it was as easily repaired. "When they spring a leak," wrote an early Canadian traveller, "they run them instantly ashore, pull them from the waters, and turn the bottom up; a fire is then kindled, and a burning cleft faggot is taken and run along the seams, while the voyageur blows through the cleft; this melts the gum, which is then pressed down by the thumb, and so the cure is effected. If a hole has been punched in the bark, the piece is extracted and a new piece inserted. When done she is soon in the water and away again on the voyage." William Armstrong painted this Indian repairer at work in 1881.

time of year was early summer, when the bark was most flexible and tough. The first step was to lay the bark in long rolls on a frame, suspended by four posts. Next the gunwales and ribs were built, usually of cedar, and sheathed with thin, flat slats of the same wood. The bark was fitted to the frame and sewn together with roots of cedar or fir soaked in hot water. Finally gum from the pitch pine was boiled and daubed onto the seams to seal them. Sometimes trees were found so large that a twenty- or twenty-five foot canoe could be built without seams from a single piece of bark.

Samuel de Champlain was one of the first white men to recognize the value of this native Canadian invention. "In the canoes of the savages," he wrote in 1603, "one can go without restraint, and quickly, everywhere, in the small as well as the large rivers." To the uninitiated it might seem a tipsy, fragile craft, but in the hands of competent boatmen the birchbark canoe was one of the most efficient methods of travel ever devised. Its light weight made it easy to manoeuvre in the water and to carry on portages, yet it could be loaded safely until its gunwales rode within a few inches of the water. If it was easily damaged by rocks or floating logs, it was almost as easily repaired or replaced. Champlain, Brébeuf, Frontenac, La Salle and all the other adventurers and missionaries relied upon it to develop and extend New France.

To the Indians the canoe was the basic means of transportation over any appreciable distance, just as the automobile is to us today; and at great tribal gatherings—or later at trading-posts or Methodist camp-meetings—a hundred or more canoes might be drawn up on the nearest

For an important European, the canoe could be a comfortable, almost elegant, form of travel through the Canadian wilderness. After one such trip Frances Hopkins painted the experience, showing herself in a Hudson's Bay Company canoe seated next to her husband Edward.

beach. The early European settlers were equally dependent upon boats. In the beginning they had no alternative to the St. Lawrence as a thoroughfare. It was some time before the first roads were blazed through the forest, and even then mud and ruts all but closed them to traffic during most of the year. The lakes and rivers were Canada's first highways. Boats of European design could be used on them where few portages were necessary; on the smaller rivers the canoe was king, and in it the early Canadians travelled widely. An English visitor in 1785 reported that the people of Quebec considered it a disgrace not to have travelled to Mackinac, or to the Grand Portage at the head of Lake Superior: as for the girls, they had no love for a man who had not made at least one such trip!

Consider the journey of a French-Canadian family which moved in the 1790s from Montreal to a new colony along the Illinois River, some 1,100 miles away. Father, mother, and four children, with all their belongings, travelled in one bark canoe fifteen feet long and three feet wide. The parents paddled at the bow and stern. The oldest child also paddled, near the centre, and the rest sat on mattresses and other baggage. During the day they stopped only once, to eat. At night they hauled the canoe onto the shore, set up a rude tent of sheets supported by two poles, cooked their supper over an open fire, then wrapped themselves up in blankets on the ground until daybreak. By 8 A.M. they were back in the canoe. Hugging the shoreline for safety, they could usually travel fifteen to twenty miles a day; but if the weather was bad, or if they encountered rapids or some other obstruction, they might lose a day or more. At that rate the trip must have taken more than two months. Yet when the family was seen near Niagara-on-the-Lake by a visitor from France, the Duc de La Rochefoucauld-Liancourt (who wrote about them in his *Travels*), they were singing.

The English settlers in Upper Canada also used canoes, although on the whole they never achieved the same proficiency as the French-Canadians, and often depended upon the latter as boatmen. To a

person of wealth or station in the colony, even a long trip by canoe could be pleasant and romantic. Anna Jameson, wife of a lawyer who for a time served as vice-chancellor of Upper Canada, travelled from Manitoulin to Penetanguishene with a brigade of canoes in 1837 and described the experience in *Winter Studies and Summer Rambles in Canada*. Her canoe was twenty-five feet long and four wide. She shared it with three other passengers, an assortment of baggage, seven French-Canadian canoemen, and in the stern an Indian steersman with a very long paddle. The *voyageurs* were colourfully dressed—a tasselled cap or a handkerchief twisted about the head, a gaudy shirt, and a pair of trousers with a gay sash and lurid garters formed their costume. The route carried the travellers through the lovely islands of Lake Huron and Georgian Bay. Each day at sunset they chose an island on which to camp and enjoyed a supper of trout or whitefish. Next morning they were off again at break of day:

> The voyageurs measure the distance by *pipes*. At the end of a certain time there is a pause, and they light their pipes and smoke for about five minutes; then the paddles go off merrily again at the rate of about fifty strokes a minute and we absolutely seem to fly over the water. "Trois pipes" are about twelve miles. . . .
>
> This day we had a most delightful run among hundreds of islands, sometimes darting through narrow rocky channels so narrow that I could not see the water on either side of the canoe; and then emerging, we glided through vast fields of white water-lilies; it was perpetual variety, perpetual beauty, perpetual delight and enchantment from hour to hour. The men sang their gay French songs, the other canoe joining in the chorus. . . . They all sing in unison, raising their voices and marking the time with their paddles. If I wished to hear "En roulant ma boule, roulant," I applied to Le Duc. Jacques excelled in "La belle rose blanche." Louis was great in "Trois canards s'en vont baignant."

Most canoe journeys, however, were arduous and demanded great hardihood and courage. The usual fur-trader's route to the Upper Lakes ran from Montreal up the Ottawa and Mattawa Rivers, across Lake Nipissing and the French River to Georgian Bay, and thence to Michilimackinac and along the northern shore of Lake Superior. The boats most often used on it were the great *canots du maître*, thirty-five to forty feet long with a crew of eight to ten men. Fully loaded, they carried three to four tons of freight: steel axes and knives, copper pots, rifles. blankets, trinkets, and, too often, rum to trade with the Indians for fur; the men's personal belongings, which were usually minimal; biscuit, pork, and pease, for there was little time to hunt or fish; and such essentials as waterproof tarpaulins, a sail, towing lines, and bailing equipment. Every item had to be carried at the portages—and on this route there were as many as thirty-six such stops.

To pack nearly two hundred pounds, often for a mile or two, was a feat at which the *voyageur* excelled. Usually he carried two bales or pieces at a time, each of ninety pounds, on a tump line, a leather strap passed across his forehead. The first was tied so that it lay on his back a little above the kidneys, the second sat upon the first, and thus laden he trotted off along the trail. On long portages deposit points were set up every two or three miles, and the carrying process was repeated in stages until the entire cargo was shifted. Then the long canoe itself was portaged on the *voyageurs'* shoulders. Except once for a meal and

In the Quetico region of northern Ontario, an ancient portage trail still can be traced winding across the forest floor.

once or twice to light a pipe, nobody stopped working until nightfall. Work days of twelve and fifteen hours were normal. No wonder that the men who made such journeys were proud and competitive, or that they considered the canoe their best friend.

> Tu es mon compagnon de voyage!
> Je veux mourir dans mon canot.

As the search for prime pelts carried the fur-traders further and further west, trade routes surmounted the height of land and penetrated the prairies. From Lake Superior to the Red River the *voyageurs* could choose from five routes, none of them easy. Only two were commonly used. The original French route, beginning at what is now Fort William, followed the Kaministiquia River for fifty-two miles and then the two Dog Lakes and the Dog River for another forty-five miles. In these two stretches alone there were fifteen points at which the journey was interrupted: some were portages, over which both goods and canoes had to be carried, others were *décharges* where only the goods had to be unloaded and carried while the emptied boats ran the rapids or shallows, were tracked through them by ropes, or were handled by men walking in the water. From there the route continued across Lac des Mille Lacs and through other lakes and rivers totalling 165 miles and nineteen portages or *décharges*. Then to Rainy Lake, Rainy River, Lake of the Woods, and the Winnipeg River to Lake Winnipeg, a grand total from Montreal of fifty-three portages and nine *décharges*. Most of the portages were less than a mile, but one was more than two and a half miles.

The early British traders had their own way of reaching Rainy Lake, following the Pigeon River and the Grand Portage. In 1796, however, the Pigeon became part of the United States, and the Kaministiquia Portage, abandoned in the course of past years, was rediscovered and came into its own again. The *voyageurs* were not unhappy, for though

Scale of Miles

0   200

FUR-TRADE ROUTES

Left: The fur-traders spanned the continent with a commercial network based entirely on the canoe. But when the water grew too rough or too shallow, the boats and their heavy cargoes had to be carried overland. The voyageurs boasted of their feats of strength on such portages. They spoke less often of the wooden crosses along the route—reminders that their life was as hazardous as it was romantic. ("Burial Place of the Voyageurs" by W. H. Bartlett.)

Water was also the key to transportation in another of Canada's early industries—the timber trade. (W. H. Bartlett: "Junction of the Ottawa and St. Lawrence.")

the Pigeon River route was 55½ miles shorter, it involved one portage twelve miles long which sometimes took seven days to complete.

The western journey was one of hardship and sometimes of privation if not death by drowning or starvation or violence. At the worst rapids there were wooden crosses marking the graves of unfortunate *voyageurs*. Yet the North West Company extended the fur trade across the continent. Sometimes horses were used on the prairies, and birch sledges and toboggans and dog-teams in winter, but the canoe was always the prime means of mobility—the *canot du maître* from Montreal to the Grand Portage, and from there westward the middle-sized *canot du nord*, which was about half the size and carried four to six men and one and a half tons of goods. The principal routes to the west from Lake Winnipeg followed the south branch of the Saskatchewan River to the Red Deer and Bow Rivers, and the north branch to Fort Edmonton. To reach the far north, canoes travelled the Saskatchewan to Pine Island Lake and thence up the Sturgeonweir River to Frog Portage; then along the rough Churchill River to Portage La Loche, over the height of land which separates the Hudson Bay and Arctic drainage basins; then the Clearwater, Athabasca, and Slave

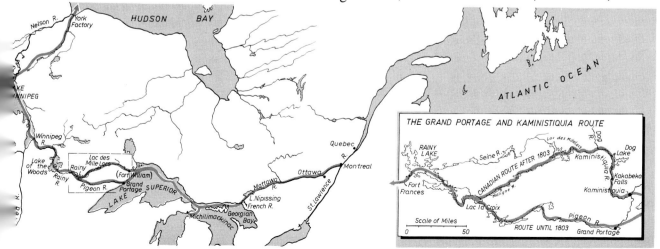

Rivers to Great Slave Lake, and on to the Mackenzie River and the Arctic. The fur-traders even penetrated British Columbia, and furs purchased from the Indians on the West Coast were shipped to Montreal by canoe in the early 1800s along this natural trans-Canada highway.

This was the high point of canoe travel in Canada: a commercial empire spreading to the western limits from the headquarters of the North West Company in Montreal. Each spring brigades of big canoes, loaded with supplies, set out along the St. Lawrence, eventually to reach the head of Lake Superior. There they met the "wintering partners," who had spent the previous months collecting furs in the west and now were waiting to turn over their pelts in return for food

and trade goods. The exchange made, the eastern brigades returned to Montreal, where the skins were shipped to Europe. The distances were both heroic and suicidal. Costs of transport rose so high that the North West Company could no longer compete with the rival Hudson's Bay Company, which shipped its goods by sea directly from London to its posts on Hudson Bay and James Bay, and from there by river into the interior. In 1821 the two companies amalgamated, and the Hudson Bay route became dominant. "York boats" instead of canoes were commonly used thereafter on the great northern rivers, even though portaging them was a back-breaking job. The old route via the Ottawa River was kept open for some years, but the great days of canoe travel were drawing to an end.

Meanwhile the canoe was giving way in the east. The French had always used sailing ships and small boats on the St. Lawrence below the rapids of Lachine, and at a quite early stage sailing vessels had been introduced on the Great Lakes. Soon flat-bottomed bateaux and the

A brigade of York boats sweeps down the Saskatchewan under full sail in this painting by Paul Kane. Boats began to replace the canoe on western rivers after the Hudson's Bay and North West companies merged in 1821. Their advantages, to Sir George Simpson, the company's general superintendent in North America, were clear-cut: "The saving in wages alone will materially exceed one-third and the property moreover will not be so liable to damage and injury on the voyage." The York boat was light enough to be portaged on rollers, and a 40-foot model could carry ten men and nearly five tons of cargo. When the square sails could not be used, the crew pulled on heavy sweeps.

larger, scow-like Durham boats were used on the upper St. Lawrence as well. By the 1830s hundreds of these craft were in operation. They did not furnish a very comfortable passage. Passengers and cargo had to be disembarked at every stretch of rapids so that the boat could be poled or towed through the shallows or lifted bodily out of the water and pulled on rollers over a portage road. For nearly two generations immigrants aided their own progress to new homes by helping to drag the boats on which they were nominally travellers; between these periods of back-breaking exertion they sat and slept crowded in the open, shivering from river-damp and fever. On the way back the boats were lighter and could run the rapids, but that was a dangerous operation calling for skilled boatmanship. Matters did not really improve until the opening of canals around the worst of the white waters.

Canals were a luxury, though, confined for many years to the relatively populous St. Lawrence Valley. Elsewhere the overland portage routes remained all-important. Their locations are remembered today in place names unchanged since pioneer times: the Pas, the Carrying Place, Rat Portage, and Rocky River. Where traffic was heaviest the paths were gradually improved. This was most evident at the principal junctions of the Great Lakes. Soon after the American Revolution the British began building a road on the western shore of Niagara Falls to parallel the old trail on the eastern—now the United States—side. By 1789 a group of traders had begun carrying loads over this route between Chippewa and Queenston at a rate of 1s. 8d. per hundred-weight. Of that amount twopence went towards maintaining the eleven-mile portage. The haulage was crude, but satisfactory for canoes, bateaux, and trade goods; and around 1800 it actually served a vessel of more than seventy-five tons. At Sault Ste Marie the North West Company cleared a forty-five-foot road which continued in heavy use even after a parallel canal was built in the 1790s. And so, at these two points and at many others along the paths of the *voyageurs*, the narrow portage trails through the woods grew into crude roads—early links in a network of highways which one day would span the continent by land instead of water.

At the Long Sault Rapids of the St. Lawrence, early immigrants to Upper Canada had to disembark and proceed by foot or wagon while their boats were hauled upstream. It was only one of many stops on the way: the trip from Montreal to Kingston took ten to fourteen days. By the time W. H. Bartlett sketched this scene in 1837 a canal was being built round the rapids, but it was not completed until 1843.

Cornelius Krieghoff: "The Ice Bridge at Longueuil."

2

Along the
St. Lawrence

THE INDIANS OF THE ST. LAWRENCE VALLEY had no knowledge of the wheel, one of man's most important inventions, nor did they have horses. The French brought both across the Atlantic—a relatively easy matter. Building the roads on which to use them proved much more difficult. For the first sixty years after the founding of Quebec in 1608, transportation was almost entirely by water. The St. Lawrence River was the original highway of New France, and its tributaries were the sideroads. The first settlers travelled over these waterways by boat in summer; in winter they followed footpaths on snowshoes through the woods along the shores. There was no time—and no need —to cut proper roads through the virgin forest. It was all the hard-pressed colonists could do to clear small plots of land and organize their tiny settlements in the face of the rigours of the climate and the attacks of the Iroquois.

There was another reason why roads were slow in appearing. They were not wanted by the men in charge. For more than half a century New France was not a colony in the usual sense. It was simply real estate, in effect leased by the King to a succession of private companies who were granted a monopoly over the valuable trade in Canadian furs. In return the companies agreed to develop the colony and carry settlers to it, a promise they did very little to keep. Quebec was founded in 1608, but the first real colonist did not arrive until 1617; no ox-plough was in use until 1628, and another seventeen years passed before the first horse was carried across the ocean to Quebec, to be presented to the governor. The companies, after all, were concerned with making money. Settlers were not profitable: on their way across the Atlantic they took valuable space which might better be filled with goods for barter, and once arrived they contributed nothing to the fur trade, which depended entirely on the trapping skills of the Indians. They might even begin buying furs on their own and cut into the profits of the company's monopoly. By 1663 there were still only about three thousand Europeans in New France, huddled close to the river in a narrow fringe east and west of Quebec City, and in occasional pockets up the St. Lawrence. Cleared land stretched almost continuously from Cap Tourmente, twenty miles below Quebec, to Cap Rouge, seven miles above it, but all the farms fronted on the river; they were long and narrow and the farmhouses stood very near to one another on the bank, so that travel by water was easy. Upriver from Cap Rouge lay only a few established farms on the large estates, or seigneuries, and clusters at Cap de la Madeleine and about the town of Trois Rivières; beyond that unbroken forests stretched to the mission and fur-trading post at Montreal.

Then the government stepped in. In 1663 New France was declared a royal province, with a governor responsible to the King and his ministers. Louis XIV poured money, men, and administrative talent into the colony, and immigration began to swell. Within a few years the population had almost doubled; by 1681 it approached ten thousand. The land along the St. Lawrence from Montreal to Quebec was gradually cleared, and a second row of farms began to appear behind the original settlements fronting on the river. To link these back lots with

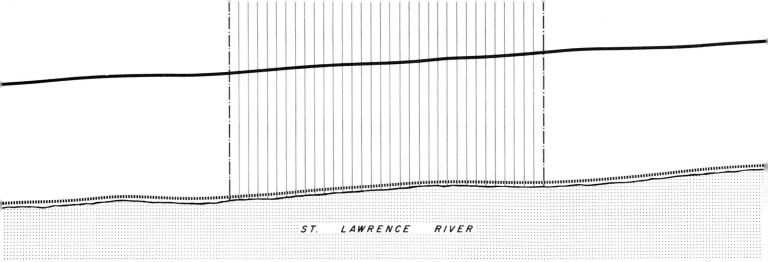

ST. LAWRENCE RIVER

the St. Lawrence the farmers themselves built the first rough roads in New France and they were obliged by law to build and maintain a road in front of the second row of farms.

New France was meanwhile receiving shipments of cattle, sheep, goats, and horses. Twenty horses were brought over in 1665; by 1685 there were 156, and after that the number grew so rapidly that Governor Vaudreuil feared his young men would lose the art of walking; he suggested killing some horses and selling them to the Indians as "beef." With draft animals at last available, the colonists began to chafe at the difficulties of water transportation: the hazards, the dependence on fair weather, the portages round rapids and falls, the interruptions during the winter freeze-up and spring thaw. Gradually roads began to appear alongside the St. Lawrence, joining with the sideroads to the interior farms. One of the first of these "highways" ran from Cap Rouge to Quebec. Part of its route is today the Grande Allée, one of Quebec's main streets.

The road network grew slowly. Comparatively few roads were built before 1700, and even after that date the rate of progress was never more than modest during the French regime; yet under the existing conditions the achievement was remarkable. Every mile involved tremendous labour. Trees had to be cut down and their stumps dragged out of the earth. Swampy sections had to be filled in, occasionally with stones but usually with logs. Except for such reinforcement, all were dirt roads. Streams and rivers abound in Quebec, and each had to be crossed somehow: if the water was shallow, the current not too strong, and the bottom firm, horses and carriages could make an easy ford; otherwise bridges had to be built of logs, and for the larger rivers ferries had to be used. Money was scarce, and so was manpower. Most of the colony's energy was absorbed in the struggle to clear the land for farming and to protect its homes from the Iroquois. Moreover many of the younger, stronger men were off in the wilderness for the greater part of the year, attracted by the adventure and wages of the fur trade.

There was no department of highways responsible for building and tending the roads, just a single official, the *grand voyer* (highway superintendent), and a few deputies. The *grand voyer* was expected to co-operate with the wishes of the residents; but he had the authority to choose routes, supervise construction and maintenance, decide where signs and ditches would be placed, and prevent encroachment by houses

After every heavy snowfall, the farmer was expected to beat down a trail along his property line so that travellers could follow a relatively firm surface. (William Cruikshank: "Breaking a Road.")

Winter travel could be dangerous. In bad weather a man could easily lose his way. Cedar trees or poles were set out to mark the safest route over rivers or lakes. On the back lines, farmers were obliged to erect similar guides along the length of their properties. (Clarence Gagnon: "Pont de glace, à Québec.")

or commercial establishments. The work was done for the most part by the local inhabitants. Under the *corvée* or "forced labour" system, each settler in effect was responsible for the road in front of his own land; but since his frontage was usually small, so was his obligation. If he had the money a man could avoid roadwork by paying a small fee. In some cases, where roads were being cut through unsettled sections between two towns, all the work was done by wage-labour.

The *grand voyers* recognized three main classes of roads. The first consisted of *chemins royaux et de poste* (King's highways), which were supposed to be twenty-four feet wide with three-foot ditches at the side. Everyone was expected to do his share in building them, from the greatest landowner to the least of his tenants. Often the work party was commanded by the local militia captain, who in general acted as the government's agent in rural districts. The captain was not necessarily a wealthy man, but normally he was one of the most respected and influential in the area, and when the roads were being built even the landlords, the seigneurs, followed his orders, and worked or paid somebody to replace them.

The other two classes of road were left to individual initiative. *Chemins de communication*, the sideroads leading to farms off the main highways, were the responsibility of the settlers who wanted them. Regulations called for them to be eighteen feet wide, but many were likely narrower. *Chemins de moulin*, of no standard size, were built at the seigneur's orders by his tenants so they could carry their grain to the mill which he was obligated to provide for them, and they to use, by the terms on which land was held under the seigneurial system.

Once the road was built the habitant was expected to help keep it in usable condition, particularly to repair damage done by spring thaws and floods. In wintertime he was also expected to beat down the snow banks and clear the worst of the ridges and ruts, and to mark the edges of the road with upright poles or small trees set in the snowbanks as a warning to travellers. These guide-posts, and the mile-posts that often accompanied them, were considered so important that a person who removed them could be sentenced to a flogging, since to wander off the road in a blizzard might mean death. Summer maintenance seems to have been much more laxly enforced. If the *grand voyer* was negligent in his visits the roads quickly degenerated into quagmires; this happened even to some royal roads a few years after their construction.

Road travel was never easy during the eighteenth century. In April hardly any stretch was usable, and at the best times the way was often rough and muddy. There was no overall plan of construction. Routes followed the meanderings of rivers, or led over hills which could have been avoided. The early road-blazers left an inheritance which has cost many millions of dollars to widen, straighten, and make more efficient. Yet narrow, rutted, boggy, and uncertain as they were, section by section the roads got built.

The most important roads in the early years were those which provided access to the town of Quebec, the colony's administrative and financial headquarters, from Beaupré, Ile d'Orléans, and other parts of the surrounding countryside. Along these roads the ox-carts of the habitants carried the farm produce to the Champlain market, and messengers rode on horseback on the business of the government and the seigneuries. Between 1709 and 1713 a route was cleared along the south bank from Lévis, opposite Quebec, more than one hundred miles downstream to Rivière du Loup, where it joined a rough portage trail all the way to the Maritimes. About the same time there was considerable activity in the Montreal area, and on both shores between Montreal and Quebec. By 1720 several sections of a road joining these two centres had been built on the south shore, but the route was not completed before the capture of Quebec in 1759. On the north shore, which was the more densely populated, the cities were finally linked in the 1730s.

Undoubtedly the work was spurred by the ambitions of a young lawyer from Paris, Eustache de Boisclerc, who in 1721 obtained a monopoly to carry mail and passengers between Quebec and Montreal —and the authority to build the necessary road. By 1734 it was possible under good conditions to travel from Quebec to Montreal by coach in four and a half days. The same trip today takes only a few hours by car, but before a road was built it sometimes took a month by water when the wind was unfavourable.

In 1745 the road eastward from Quebec was extended with great difficulty to Baie St. Paul, but it was eventually abandoned and re-located in 1806. On the south shore the roads began to move away from the St. Lawrence. The road from Fort Chambly to Montreal dates from 1739; that from Fort St. Jean dates from 1749. In the 1740s, too, roads were opened in the area south of Lévis, and in 1758 the Route Justinienne extended all the way to the Maine border through the Beauce.

The most common vehicle in New France was the *calèche*, a sturdy two-wheeled gig the settlers had brought with them from France. It could travel on almost any road, and unlike a four-wheeled carriage it could not easily be upset. The *calèche* had room for two passengers on a high seat which was mounted on "grasshopper springs," according to one authority, who added: "The driver occupies the site of the dash-board, with his feet on the shafts and in close proximity to the horse, with which he maintains a confidential conversation throughout the journey, alternately complimenting and upbraiding him." It is still possible to take a short tourist ride in a *calèche* in Quebec City today,

The most common vehicle on the roads of Lower Canada was the calèche, *a sturdy two-wheeled gig with a high seat mounted on "grasshopper springs." According to Isaac Weld (who made the original drawing from which this etching was taken), the drivers used only three expressions in urging their horses onwards—"Marche!" "Marche donc!" and a piercing "Marche donc!!" accompanied by the whip. As a consequence* marche-donc *was a popular nickname for the* calèche.

but in early times they were used for long trips, though their open sides offered no protection from rain or mud, and their high bodies must have swayed uncomfortably over the bumpy roads.

In winter the *calèche* was replaced by the carriole, a kind of sleigh which was basically a cart frame mounted on iron-shod runners. Carrioles came in a great variety of shapes and sizes. Some were narrow and looked like coffins, others were wide and raised on runners like a cutter. Most were pulled by horses, but sometimes dogs were used to draw children's carrioles. In some of the carrioles the passenger drove himself as he sat muffled in fur robes; in others a driver stood at the front with two passengers behind him. Where there were no roads, the carriole could be taken as easily over the frozen lakes and rivers, and usually the rough ice was no bumpier than the highways. Certainly nothing could be more pleasant for a young man and his "muffin" (girl-friend) than dashing along cosily packed in their gay carriole!

New France had neither enough people to require an extensive system of roads nor the economy to build them. Though immigration had risen with the advent of royal government, newcomers still only

trickled into the vast area of the colony; the French did not cross the Atlantic in the same numbers as the British who, fleeing religious persecution, filled the colonies on the Atlantic seaboard of the United States. The Canadian population under French rule never exceeded seventy thousand persons, a large number of whom lived in Quebec, Montreal, and Trois Rivières. The rest were spread out thinly along the St. Lawrence and up the Ottawa and Richelieu Rivers, locked in the valleys between the rocky Laurentian Highlands to the north and the Appalachians to the south and east. There was no heavy industry to require roads—only shipbuilding, which of necessity was beside water, and some early iron forges along the St. Maurice River near Trois Rivières. Farmers close to the larger towns needed roads to carry

their produce to market, but in general the settlers were self-reliant. They grew their own food, spun and wove their own clothing, and milled their own grain.

The fall of New France made little immediate difference. The *corvée* system was retained, although the habitant, never anxious to leave his own fields to work on the road, was even more reluctant to do so under English orders. The British also kept the office of *grand voyer*, but for the first twenty years he had little to do: all available manpower was needed to meet the threat of war with the American colonies. In 1783, however, work began on a new route to the Atlantic that would avoid United States territory, and three years later this coach road was opened all the way to New Brunswick via the Temiscouata Portage. During the next two decades further roads were pushed south and eastward to help develop the south shore and to communicate with the United States. In 1792 a nine-foot-wide road was extended from Trois-Pistoles to Pointe-au-Père, beyond Rimouski. Other roads were opened into the Beauce. Craig's Road, which cut diagonally across the Eastern Townships from Quebec to Richmond and then south into Vermont, was planned in 1800 and completed as far as Shipton by 1810; a regular stagecoach service from Quebec to Boston began the following January.

On the north shore the roads were slower in spreading into the interior. There was no passable road from Montreal to Hull until the late 1820s, for example, and travellers used the ice of the Ottawa River in winter. In many other areas, too, the frozen lakes and rivers remained the best highways available. They were so popular that in more thinly settled regions innkeepers, and often farmers, set up temporary taverns on the ice.

> Sometimes [wrote John Mactaggart] they will remain too long in these inns after the thaw comes on, being greedy and not removing their quarters so long as they are catching a farthing; floods will therefore come on, sometimes during the night, and sweep all to desolation. . . . Whole families have thus been hurried away and drowned; and others brought out of their floating houses alive after drifting many miles down the rivers.

At Quebec, ice was the only possible bridge across the St. Lawrence to the southern shore. The passage to Lévis was seldom completely frozen over—perhaps about once in six or seven years—but when the

In eighteenth-century Canada many loads were still carried short distances on the backs of men. Heavier goods were hauled by horses or oxen. This engraving, published in 1761, shows the ruins left after the bombardment of Quebec by the British fleet. (Richard Short: "A View of the Bishop's House with the Ruins As They Appear in Going down the Hill from the Upper to the Lower Town.")

Winter or summer, the market places of Quebec City were filled with travellers. One of the principal reasons for building early roads was to provide a means for farmers to carry their produce to the city for sale. Both these paintings were done by J. P. Cockburn.

shout arose: "Le pont est pris!" people dashed out in horse-drawn sleighs to enjoy the treat and to sample the wares of the shops and taverns hastily set up on the river. At all other times the journey had to be made by boat—an easy crossing in summer but sensational when the winter ice broke up and the hardy canoemen fought their way across, carrying their birchbark craft over the pack in one place, dragging it in another, then launching it into the freezing water and paddling for dear life among the floes. The mail, among other things, was carried in this way, to the astonishment of tourists. Mrs. John Graves Simcoe wrote in her diary that such travel was "hardly credible till it has been seen."

Mrs. Simcoe, wife of the first lieutenant-governor of Upper Canada, also recorded her impression of the carrioles, which were "driven furiously, as the Canadians usually do," over ice so rough it made her quite seasick. The streets of the town of Quebec, she noted, were just as rough, and so full of holes and pits that travel on them was "very jolty." This was in the winter of 1791–92. In December Colonel Simcoe travelled to Montreal from Quebec in two days: the first stretch to Trois Rivières by carriole, and the remainder by *calèche* after crossing the St. Maurice River by boat. He was more fortunate than the summer traveller. The Quebec-Montreal road, although the most important in the province, was full of bogs, ruts, and rocks. There were no drainage ditches. The bridges, constructed of rows of poles and hardly wider than a *calèche*, were dangerous traps. When the water was high the bridge poles often floated.

Several laws were passed between 1780 and 1819 to facilitate travel along this route. The sixty leagues were divided into twenty-four stages, and the postmasters were obliged to keep four *calèches* and four carrioles ready at a quarter hour's notice to forward passengers from stage to stage. To the traveller each stage was an opportunity to step down and stretch his limbs, and perhaps have some refreshment while he waited for the fresh horses or conveyance. The arrangements were made with no little courtesy, and when friends passed through there was a great deal of kissing and *bon voyage*ing in all directions.

Besides his carriage trade each postmaster provided saddle horses for the mail couriers. These express riders regularly travelled 180 miles in thirty hours on a weekly service. Their route was subsidized to some extent: the postmasters rented horses to them at 6*d.* a league, half the passenger rate, and the ferrymen transported them across rivers free of charge. Mail was also carried by riverboat captains, who received one cent for each letter they delivered along their routes. As stage service improved, these older forms of postal service were gradually replaced.

The British regime gradually modified the French system of road administration to make it more democratic, placing it under the supervision first of the Legislative Council and then of the elected Assembly, and providing for the election of some highway officials in the parishes. Increasingly it became evident that a major change was necessary. The *corvée* system was too inefficient. It depended on the labour and desire for roads of habitants who would much rather be working on their own

The frozen river provided a favourite outing for the young people of Quebec and Montreal. "On the fine, frosty moonlight nights, when the sleighbells ring merrily and the crisp snow crackles under the horse's feet," wrote Isabella Bishop, a visiting Englishwoman, "the gentlemen call to take their 'muffins' to meetings of the sleighing-clubs, or to snowshoe picnics, or to champagne suppers on the ice, from which they do not return till two in the morning." The painting by R. G. A. Levinge shows Quebec City and the St. Charles River in 1838.

When the St. Lawrence River froze solidly from Quebec City to Lévis, townspeople from both sides met in a temporary village of stores and huts set up on the ice. J. P. Cockburn, then a young army officer, recorded the scene in 1831.

farms, and it simply was not designed for costly construction projects like big bridges.

Several other systems were tried. In 1806 the government began receiving tenders for road construction, but since no taxes were voted for this purpose the contractors were paid in land instead of money. Few contractors, it seems, were interested in working under these conditions. To build the road which bears his name, Lieutenant-Governor Sir James Craig sent 180 British soldiers to clear the way, but such methods never became common practice in Lower Canada.

A far more successful solution was the toll road. Travellers paid a fee each time they used these roads, and the problem of financing therefore was transferred from the reluctant landowner to the man who was directly benefited. The first toll or turnpike road was established in 1805, but this innovation was not followed widely until after the War of 1812. Then petitions flowed into the Legislative Assembly requesting the construction of new toll roads, and for the rest of the nineteenth century this method of financing met the needs for long-distance travel without the aid of additional taxes.

But toll roads did not completely replace the public road system. Some roads in fact were free along part of their length, and were turnpiked along other stretches. In 1815 the government moved towards the system we know today. It set aside specific funds to be used for road and bridge construction by professional contractors. Each county and district received a set amount, to be spent under the control of three road commissioners. The sums were not large—roads rarely received more than a minimal proportion of public money in early Canada—but they were large enough to ensure that several new major roads were built. The route from Baie des Chaleurs was extended to Percé, opening up the south shore of the Gaspé Peninsula to settlement. From Montreal a much-needed link with New York was built along the Richelieu River to Lake Champlain, and the route from Quebec to Boston down the Chaudière Valley was improved so that it could be used by carriages. Trunk roads through the Eastern Townships were expanded. By the mid-nineteenth century, as the needs of the farming communities on the south bank of the St. Lawrence were gradually met, new roads were extended inland into the north-shore frontier of the Saguenay.

The toll collector, emerging from his shelter to halt the passers-by, became a familiar figure in Lower Canada after about 1815. The signboard on the front of his house listed the fees travellers were expected to pay for the privilege of using the "improved" road.

"The Fiend of the Road."
(A Currier & Ives print
from a painting by
Scott Leighton)

3

The Atlantic Provinces

Even close to the capital, Newfoundland's roads were little better than trails in 1872, when this engraving was made to show "the road leading to O'Brien's Bridge, looking south towards St. John's."

IN THE ATLANTIC PROVINCES, as in New France, roads were slow to appear. At first the settlers looked outward to the sea from which they drew their wealth, not to the untamed land which lay behind them. Their fishing grounds, particularly the Grand Banks of Newfoundland, were the richest that man has ever known, swarming with cod in such numbers that in the early days no lines were needed—the fish could be caught in baskets. The ocean also provided the route to markets in Europe and the West Indies where the dried and salted cod were sold. Within the colonies the sea offered the natural and for a long time the only means of travel between communities. In scores of inlets along the rocky coasts, villages grew up out of contact with the interior. Some, like Lunenburg and Liverpool, were prosperous before they had any means of land communication with the rest of the province. The route by sea was roundabout and often stormy, but it already existed and to use it cost nothing, whereas roads were difficult to build and expensive to maintain.

Newfoundland was the first of the Atlantic provinces to be discovered by Europeans, and the last to have roads. More than four hundred years ago its bays and coves were already well known to European sailors who crossed the Atlantic to fish for cod, but for a very long time the island was little more to them than a mother ship conveniently stationed near the Grand Banks. They landed on it to find wood and water; in its harbours they built wooden platforms on which to dry and salt their catch; and at the end of the summer season they went home. Only a few remained. Several attempts were made to establish a colony, beginning with Sir Humphrey Gilbert's expedition in 1583, but invariably they failed. The land was inhospitable, and at home powerful interests fought to prevent any permanent settlement for fear that resident fishermen would cut into the profits of the vessels that sailed each year from England. Indeed, the inhabitants were not legally entitled to construct permanent dwellings until 1811. During the 1700s, moreover, settlement was discouraged by war between the English and the French, both anxious to control the fishing grounds.

Some trail-blazing was done, but not by the island's few resident fishermen. The men who cut the first paths had no access to the sea—in fact they were forced to hide in the woods to escape the law. They belonged to a band called the Society of Masterless Men. Most of these outlaws had come to Newfoundland as impressed sailors, forced into the service against their will, or as indentured servants, kidnapped from Ireland and sold to fishing-masters and landowners at so much per head. Having escaped from the cruel laws of the day, they lived in the forest by hunting caribou, by raiding government stores and wealthy farms, and by some stealthy trading. With the support of the working people they grew in numbers and strength during the last decades of the eighteenth century; and over the years they built a network of paths across the Avalon Peninsula. Some of their more important routes still exist as wood trails, used between villages until the present century. Others led nowhere: cut to confuse pursuers, they petered out in bush or bog. The men of the Society never were mastered. Eventually, as

the laws against private settlement were eased, they married and began new lives along the coast.

It was not until the mid-eighteenth century that Newfoundland's permanent population began to grow to sizeable proportions. Even then it was widely scattered outside the capital at St. John's, in tiny, isolated villages of perhaps ten or twenty unpainted wooden shanties along the shores of deep, rocky fjords. In St. John's the principal streets were already being laid out in 1773, but half a century passed before they were extended beyond the town. The first true road was cut westward eight miles across a peninsula to Portugal Cove on Conception Bay. During the governorship of Sir Thomas Cochrane from 1825 to 1834, a few other settlements up and down the Avalon Peninsula were linked by road to the capital, among them Bay Bulls to the south and Torbay to the north. Further development was slow. Much of the government revenue was swallowed by the salaries of officials, with but little spent on roads or other improvements. In the late 1840s a "main road" in the colony was proposed, and a select committee suggested that £100,000 be borrowed in England for the purpose, but not much came of the project. In fact, many areas of Newfoundland have been joined by road only since its confederation with Canada. Hundreds

Robert E. Holloway, photographing Newfoundland late in the last century, found only the roughest of tracks leading from North Shore outports like Moreton's Harbour on Notre Dame Bay. It was such conditions and worse that led Sir Richard Bonnycastle to write in 1842: "Open then the roads to the distant settlements, and set the inhabitant of the woods and iron bounded shores free from the moral and intellectual bondage he now labours under, for want, in many places of rudiments of religion, and of the commonest education."

of bays, coves, and inlets make the six-thousand-mile shoreline so irregular that road building along it has not always been feasible or possible, and until quite recently the interior was so sparsely populated that it required no road system.

In the 1850s, certainly, there was little enough in Newfoundland to justify expenditures for roads. No industry but fishing was permitted. There was not any farming to speak of: the villagers occasionally cultivated small vegetable plots behind their homes, but most food had to be imported by sea and could be purchased at the main ports when the fishing catch was delivered. The chief requirement for overland travel arose from the need to deliver the mail. Without letters and newspapers, in the days before radio, the outports could be cut off for months from all news of the outside world. To ensure that this would not happen a surprisingly extensive service developed by the middle of the nineteenth century to carry the mail up the east coast, where most of the people lived. The route followed was partly by land, partly by water, in an attempt to keep to a more or less straight line cutting across the deep bays and narrow peninsulas. From the main post office at St. John's to the town of Bonavista a letter would travel on horseback along the colony's first road to Portugal Cove; by sailing-ship across Conception Bay to Brigus, Harbour Grace, and Carbonear; then by another overland messenger to Hearts Content on Trinity Bay; by a second ship across that bay to Trinity and Catalina; and a final horseback ride across the Bonavista peninsula to its destination. From that point this peculiar land-sea service continued up the coast to Greenspond, King's Cove, Cat Harbour, Fogo, and Twillingate. Southward from St. John's the mail service extended around the Avalon peninsula and on to the Burin peninsula. In winter the trip had to be made entirely by land, and some couriers undertook to carry the mail under the greatest difficulties.

One other group did much to maintain contact with the remote settlements at this period and later. Methodist and other evangelical ministers did not wait for good roads. Some took to boats, visiting in a single season fifty or more harbours; at several they reported that they found not a single Bible, nor for that matter anyone who could read one. Others made their way by land through the summer forests, climbing hills, wading streams, plunging through marshes, with their belongings carried on a stick over their shoulders. "In travelling we never wore boots," wrote one Methodist preacher, "for they were useless to keep us dry, as we would always get over the top of any boots we could put on when wading over the numerous brooks."

On the Atlantic mainland Canada's first road (to use the term loosely) was built at Port Royal. This idyllic French colony on the Annapolis Basin was founded in 1605 by Samuel de Champlain and some associates. Food, brought from overseas and supplemented by game and the produce of gardens and fields they planted, was so plentiful that the colonists had little to do during the winter. In their spacious log quarters they created an Order of Good Cheer to help pass the time, and celebrated daily over groaning tables. Simply to avoid idleness

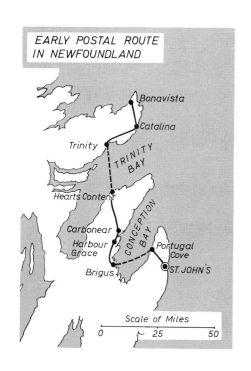

EARLY POSTAL ROUTE
IN NEWFOUNDLAND

Bonavista
Catalina
Trinity
TRINITY BAY
Hearts Content
CONCEPTION BAY
Carbonear
Harbour Grace
Portugal Cove
Brigus
ST. JOHN'S

Scale of Miles
0   25   50

Champlain decided to cut a road some two thousand paces long from the colony to a little creek so well stocked with fish that it was known as the Troutery. With the help of two or three other men, Champlain completed the path in a short time. Thereafter, the gentlemen of the Order worked off their dinner with a stroll along its length. That was in 1606, two years before Champlain founded Quebec. It was perhaps the last time that any Canadian settler would cut a road just for sport.

Nearly a century and a half passed before any significant road was built in Nova Scotia. In the interval the Acadian French extended their colony eastward from Port Royal along the Annapolis Valley and Minas Basin. Probably they built paths along the dikes and cart tracks between the farms in this area, but to travel any distance away from the seashore demanded knowledge of the woods, through which well-established Indian trails led to the south shore, the Strait of Canso, and Northumberland Strait. Then in the middle of the eighteenth century three events took place which led to more rapid development. First, in 1749 the British founded Halifax to counter the French stronghold at Louisbourg on Cape Breton Island, and a few years later they encouraged further settlement at Lunenburg. Thus two new centres of population sprang up on the south shore. Second, the French Acadian settlers were expelled from their farms in 1755 when they refused to swear allegiance to Britain. Their land was taken over by New Englanders, who were more active in their demand for representative government and good roads. And third, the end of the Seven Years' War in 1763 resolved the long-standing dispute between France and Britain over the Maritimes, which meant that the area could develop in peace.

In the same year that Halifax was founded, an eighteen-foot road was cut across the peninsula from the new capital to the old Acadian settlement at Piziquid, which, renamed Windsor, became an important British community. By the time of the expulsion of the Acadians the route had been extended through rich farming country to Annapolis. A move to link Halifax and the thriving settlement at Lunenburg in 1757 had to be dropped, however, because of the cost—roughly £6 per mile. Throughout the 1760s trails were cut to a number of outlying settlements. In the more thickly populated agricultural areas cart roads were opened for local use, usually along the lines of the old French trails. There appear to have been several county roads leading to Yarmouth, for example, long before there was any land communication with other parts of the province.

The word "road" in those days could mean almost anything. The streets of Halifax were mires, and outside the town even the most important routes were allowed to become overgrown, or blocked by windfalls and broken bridges. In the mid-1760s only one road, that between Halifax and Windsor, was fit to be used by wheeled vehicles. The way to Annapolis was reported in 1763 to be "embarrassed with young elders and brush" and in need of three new bridges, but it was agreed that thirty men working for twenty days could make it safe for horses and cattle. A year later it was described as adequate for horsemen, but still unsuitable for women. Conditions were so bad that it was cheaper to transport farm produce from Minas Basin to Halifax

A principal intersection in Halifax about 1760: the Governor's house and St. Mather's Meeting House fronted on Hollis Street (at left); up George Street (at right) can be seen the Parade and Citadel Hill. (Engraved from a sketch by Richard Short.)

three-hundred-odd miles by sea around the peninsula, than to carry it sixty miles overland. Many New England sailing captains made a handsome profit from this trade.

By 1776 the road to Annapolis had improved enough for a regular postal service to begin. The postman made the trip from Halifax every two weeks. Several hundred pounds were spent in the following years on roads in the Annapolis area, but in other parts of the colony the roads developed slowly, if at all. Often they were begun haphazardly by people in the neighbourhood, and decades might pass before the government stepped in to carry their construction further. A good example is the road from Truro to Pictou. In 1767 a handful of young men in Truro heard that a group of settlers had landed at Pictou, thirty-five miles away, and decided to welcome them. They cut a trail through the woods, stopping at one or two points to name landmarks after themselves. The next spring the Pictou settlers walked along the trail to Truro for supplies, a three-day trip each way. But no money seems to have been spent on the road until 1785, when £200 was allotted for repairs. Even in 1792 most farm produce was still being carried from Pictou by American ships rather than by road. The courier who delivered the mail to Pictou found so few settlers along the route that he habitually carried a wallet full of oatmeal, which, mixed with water from the many streams he had to cross, sustained him in his passage.

The Maritimes faced the same shortages as New France—men and money. Settlers along the Atlantic, as along the St. Lawrence, were ordered by law to help build their own roads. In 1761, when responsibility for roads was transferred from the governor to the elected General Assembly, a statute was passed requiring every man to assist in the building of new roads or the repairing of old ones; four days a year were demanded of him if he could supply a cart and a team of horses or oxen, six days if he could not. He was also expected to supply up to one day's work after a heavy snowfall. The arrangement was not equitable: payment of four shillings was accepted in lieu of a day's work, and the statute made no distinction between large and small landowners—both worked the same number of days, although obviously one was served by a longer stretch of road than the other. Moreover, for decades it missed the non-resident landowner entirely. Nor were the results very efficient: the plan had the apparent advantage of saving money; yet in fact it was wildly extravagant. The men were forced to leave productive work on their farms, and they bitterly resented it. We know of one case where a shoemaker, Matthew Wallace, threatened to cut off the head of a commissioner who had ordered him to turn out or pay a fine. Moreover, the law said nothing about how much had to be *achieved* during a day's statute labour, and as a result the men rarely did as much work as could have been bought with even moderate taxes. Statute labour had many nicknames in this period. In the United States it was popularly known as "working out the road tax," but in Nova Scotia it was "doin' a little soddin' "—for most of the repairs seemed to consist of little more than pitching earth or sod from ditch to roadway; in the words of one American commentator it was merely "a ridiculous frolic of a number of idlers." At times soldiers were ordered out to build new roads, particularly in thinly populated districts, and increasingly the government turned to private contractors to get the work done.

Road building was always expensive in the Maritimes, even when labour costs were not high, because of the many rivers and streams which had to be crossed. Some ingenious methods were used to raise the needed money. One governor in the 1760s proposed financing new roads by mining Cape Breton coal and selling it under a state monopoly. At first the British government refused permission, but later it relented and in one year £500 was received from merchants leasing the mining rights. In 1765 a special road tax was imposed on all taverns and shops licensed to sell liquor—of which there were not a few. That still did not bring in enough money, however, and seven years later taxes on land were introduced. From time to time special lotteries were held to raise funds for large projects such as bridges. The cost of beginning work on a new road to link some of the scattered settlements on the South Shore with Halifax gives some idea of the sums involved in road-building. In the 1870s more than £1,000 was voted for the project. This was an enormous amount for those days, but it paid for little more than blazing a trail and did not permit the building of bridges.

The end of the American Revolution in 1783 brought thousands of United Empire Loyalists to the Atlantic provinces. These were men and women who, either through personal choice or from force of circumstances, had decided to live under the British crown rather than in the new republic. Many put down roots in the long-settled peninsula, but some fourteen thousand others crossed the Bay of Fundy and established themselves in the St. John River valley. Largely as a result of its increased population, the Atlantic region was split into four provinces in 1784: Nova Scotia, New Brunswick, Prince Edward Island, and Cape Breton Island (which rejoined Nova Scotia in 1820).

The arrival of the Loyalists in Nova Scotia was followed by an increase in petitions to the government for new roads, improvements, repairs to bridges, winter maintenance, and the raising of standards to permit carriage travel across the province. People from the older colonies of the United States must indeed have been dismayed at the state of the highways. After a trip on horseback from Halifax to Truro in 1786, one traveller noted:

There was something like a road eleven miles from Halifax, but beyond that there was only a narrow avenue through the woods, on which the trees had been cut down and sometimes cut across and rolled to one side. The road was generally so soft that even in mid-summer the horses sank to their knees in mud and water, and as each horse put his foot where his predecessor had, the path became a regular succession of deep holes, such as one may see in a road made in deep snow.

Government grants had already been made in the previous two years to improve this particular route, and more were to be made in 1794 and 1796, but even then much of the road was described as "little more than a blazed trail." Yet it was in fairly heavy use by the standards of the time. One man kept a record of traffic passing through Shubenacadie on its way along this road to Halifax market, and in 1795 he counted 786 fat cattle, 30 cows and calves, and hogs and sheep without number; from January 20 to March 1 alone, he estimated, the farmers had carried three tons of butter, poultry, meat, and other commodities to the capital for sale. Beyond Truro the roads were in such bad repair that for years farmers in that area had no alternative but to drive their cattle to Parrsboro, on Minas Basin, and take the expensive ferry trip across to Windsor where the best highway in the province led to Halifax.

Driving cattle does not require the same quality of road as carriage travel, but in a country cut by rivers and streams which are too swift or deep to be forded, it does require adequate bridges. And as long as bridges were made of logs or planks they soon rotted and broke. There are constant references in the records of this period to roads closed to traffic because of the failure of bridges. In many places ferries were used instead, but these were not always dependable. Even the most intrepid traveller could be halted. The Reverend Jacob Bailey complained of walking in 1785 along a road knee-deep in mud in order to perform a marriage, only to find his way blocked by a river with no boat to cross it. The upshot was that the lovers remained un-united and Bailey had to return home without his fee.

Heroic efforts were made to cut new roads through the interior, and grants were offered to encourage pioneers to settle along them. Yet often within a few years the routes were overgrown, and within a generation they had vanished completely.

Human nature posed additional problems. The statute labour system, as we have seen, was ineffective. So was the system of appointing local road commissioners each spring: the new commissioner would be just beginning to learn what improvements were needed, and would perhaps be making the necessary arrangements, when his term would run out and he would too often be replaced by someone with completely different views. Graft and corruption were not uncommon. Moreover, there was an underlying battle between the House of Assembly and the governor and his council. The House, particularly those members from farming areas, were interested in roads and were urged by their constituents to provide as much money as possible for this purpose. Roads provided the easiest way, and often the only way, to get crops and cattle to market. The council, on the other hand, consisted of men who lived in Halifax; they were appointed by the governor, not elected by a populace demanding roads. They did not need rural roads; instead they were anxious to secure revenue for official salaries and public buildings. Thus every year the House voted a much larger sum for road construction and maintenance than the council was prepared to allow. Through the early years of the nineteenth century the passage of road grants was attended by constant battle, often waged with great bitterness.

The situation improved somewhat after 1808. In that year, a quarter of a century after the arrival of the Loyalists, the Maritimes received another unexpected economic boost as a result of United States policy. The government in Washington, reacting to the Napoleonic wars, closed its harbours to trade with Europe, and by the time this policy was rescinded the old commercial patterns had been disrupted. At the same time, Britain began importing lumber from the Maritimes in considerable amounts since supplies from Europe were cut off by Napoleon. By the time war ended in 1815 the harbours and towns of the Atlantic provinces were among the busiest on the Atlantic seaboard.

More money was now available for roads. Development admittedly was slow and patchy in many areas, with long stretches of road still unsuitable for anything but foot travel. As late as 1822 the route from Shelburne to Clyde River (now part of Highway 3) was described in many places as barely perceptible: "The traveller is guided by the stumps of decayed trees being marked with red paint to guide his footsteps." Elsewhere at the crossroads horsemen had to dismount to clear away fallen branches or disentangle their animals' hooves from root masses. Yet we must remember that travellers' accounts are usually weighed on the critical side, for a good road seldom draws comment. In 1823 Thomas Haliburton (who is best known for his books about the Yankee clock-maker, Sam Slick) was able to write that the two main roads in Nova Scotia, from Halifax to Pictou and New Brunswick, and from Halifax to Annapolis and Yarmouth, were "in as good condition as the best in the United States of the same length" and were being steadily improved, while the lesser roads were also "in a situation far beyond the age and wealth of the country." By the 1830s adequate roads extended westward from Halifax along the south shore to Shelburne, Yarmouth, and Digby, and eastward to Canso and Antigonish. Along many of these, stages ran regularly. Progress, at last, was surprisingly swift, as Joseph Howe described in 1837:

> Had anybody told them ten years ago, when Hamilton used to carry the mail on horseback from Halifax to Annapolis, and sometimes in a little cart with a solitary passenger beside him—who looked as if he were going to the end of the world and expected to pay accordingly—that they would have lived to see a stagecoach, drawn by four horses, running three times a week in the same road . . .
>
> Would they have believed, had they travelled on foot with Stewart, the old postman from Pictou to Halifax, who used to carry the mail in his jacket pocket, and a gun to shoot partridge for sale as he went along, that before their heads were cold they would travel between the same places in a coach and four, with a ton of letters and papers strapped on before and behind?

Howe, who was campaigning at the time for the establishment of railroads, went on to predict that the progress of the iron horse could be just as quick.

Roads on Cape Breton Island rather lagged behind those on the mainland. The first road was built by the French near Louisbourg, but it earned the governor nothing more than a stern rebuke, for officials in Paris feared it might be used by the British in an attack. By 1800 that road had disappeared in the underbrush, and there were no more than ten miles of passable roads elsewhere on the entire island. Government revenue was being eaten up by official salaries. A quarter of a century later the mail was still being carried on horseback along rough, narrow roads. It was a long time before these routes were cleared of stumps; when wagons first came into use, notches were often cut into the largest stumps to let the wheel hubs pass through them. It was not until after 1850 that any extensive road system was developed.

On Prince Edward Island, as in the other Atlantic provinces, transportation long depended on water. Although the island was settled by the Acadians early in the seventeenth century, no determined effort seems to have been made to build roads until Walter Patterson arrived as governor in 1770. He found it took two weeks to travel from Charlottetown to Princetown (Malpeque) by a combination of water and land—by boat up the Hillsborough River, then along a trail to St. Peters, and then again by boat along the coast—and in 1771 he had his troops open a thirty-three-mile road to replace the old route. He had a second road built to connect the headwaters of the Hillsborough to those of the Montague, and thence to Georgetown. Both were primitive, but the governor showed his enthusiasm by driving the first four-wheeled vehicle over them. Another member of the party said he would never forget the difficulties of that short trip; but Patterson himself believed so firmly in roads that at times he personally provided the funds to build them.

In 1774 he proposed that the settlers help build the roads under a system of statute labour. At one point every man between the ages of sixteen and sixty, except ferrymen and slaves, was required to work on the roads thirty-two hours, or four days, each July, but that regulation was so unpopular that it was quickly repealed. Statute labour continued in a modified form for nearly 175 years on Prince Edward Island, playing a basic though uncertain role in road development. As in the other provinces, statute labourers were given no expert guidance except by the surveyor who staked out the route's centre line; in other regards they were expected to use such practical skills as they had gained from farming.

New Brunswick, the last of the Atlantic provinces to be settled, followed the same general pattern of road development we have seen elsewhere, but at a greatly accelerated rate. For some time the St. John River and its tributaries were adequate for most transportation needs, and the first roads were cut along, or near, their banks. Travel upon them was anything but comfortable. In 1803 the normal method was still by horseback, though one visitor described the province as "a hell for horses."

At first road construction in New Brunswick depended entirely on statute labour, but as early as 1801 its government began making regular annual grants for road construction and maintenance, leading Nova Scotia by fourteen years. The first important road followed the St. John River from Fredericton down to its mouth on the Bay of

While the land was still being cleared in 1836, Stanley, in central New Brunswick, already boasted a large log tavern for travellers. But the grandness of the accommodation may have been exaggerated, for this sketch by P. Harry was published by a land company anxious to attract English immigrants. The crude ox-drawn vehicle, with its solid birch wheels and strong axle and pole designed for rough ground, probably gives a truer picture of conditions.

Fundy. This was a roundabout route, and in 1826 a more direct road was cut overland from Fredericton to Saint John. Another was pushed northward from Fredericton up the Miramichi to the Gulf of St. Lawrence, and during the 1820s a third principal road ran down the east coast from Chatham to Shediac. By 1827 the capital was also linked to St. Andrews, Halifax, and Quebec. Only some sections of these roads could be used by carriages, and for much of the year they were entirely impassable, but the fact remains that land communications had been developed fairly quickly throughout most of the settled portions of the province.

By 1849 New Brunswick boasted 1,269 miles of major roads between the principal centres of population. In addition there were several hundred miles of by-roads, built by statute labour and not so well planned or constructed, but joining individual settlements and even houses. The road system had cost the province £150,000 of which a

The Royal Road, New Brunswick, in the mid-1830s. The great elm tree was known as "Sir Archibald's Walking Stick" in honour of the governor, Sir Archibald Campbell. (Sketch by W. P. Kay, published by the New Brunswick and Nova Scotia Land Company, 1836.)

"A Winter Scene in Fredericton." (Lithographed in 1836 after a sketch by W. P. Kay.)

good part was spent on bridges, and for some years £10,000 a year was required to maintain it. But the trunk routes were established; all that remained was to fill in gaps, improve the grades and surfaces, and build better bridges. Thanks to its rich timber stands New Brunswick had the money to invest in good roads. The pace of development was quickening.

Of all the roads in this area, one of the most interesting and demanding was Canada's first interprovincial highway, linking the Maritimes with Lower and Upper Canada. It was first of all important as a military route which avoided United States territory. In 1792 Mrs. Simcoe met some British officers who had travelled from Fredericton to Quebec City, about 370 miles, in mid-winter. They had walked two-thirds of the distance on snowshoes; that part of the trip took nineteen days, during which, as Mrs. Simcoe recorded in her diary, their routine was

to set out at daybreak, walk till twelve, when they stood ten minutes (not longer, because of the cold) to eat. They then resumed walking till half past four, when they chose a spot where there was good firewood to encamp. Half the party (which consisted of 12) began felling wood; the rest dug away the snow till they had made a pit many feet in circumference, in which the fire was to be made. They cut cedar and pine branches, laid a blanket on them, and wrapping themselves in another, found it sufficiently warm, with their feet close to a large fire which was kept up all night. . . . One of the attendants, a Frenchman used to the mode of travelling, carried 60 lbs. weight and outwalked them all. They steered by the sun, a river, and a pocket compass.

The route from Fredericton to Quebec was known as the Temiscouata Portage, or simply the "French path." It ran up the St. John River Valley as far as Grand Falls. From there a trail led to the Madawaska River. Travellers continued down the river, which crosses the New Brunswick-Quebec border, to Lake Temiscouata, and across the length of that narrow, snaking body of water. At the north end of the lake another trail led to the St. Lawrence River, where they picked up the south shore road to Lévis and Quebec.

The Temiscouata Trail was also important as a postal route to Quebec when the St. Lawrence River was frozen over. In winter letters from Britain to Lower Canada were dropped off at Halifax. From there the trip to Fredericton was relatively easy: across the Nova Scotia peninsula by road to Windsor and Annapolis; across the Bay of Fundy by boat to Saint John; from there by land or water to Fredericton. From that point they were carried by Indians or Métis along the overland trail. The route from New York to Quebec was shorter and less arduous, but charges for passage through American territory were high.

For decades this connection between the Maritimes and Quebec, 627 miles long, remained a tenuous one, to be negotiated only at the expense of considerable time and money and by the use of ingenious expedients. George (later Sir George) Head, made the trip in 1814 on his way to a new military posting on the Great Lakes, and left a detailed account. He paid £20 ($80) for a sleigh to Annapolis, 132 miles distant; the fee covered the cost of the trip and the driver's return to Halifax. Every twelve or fifteen miles along this stretch he found inns with generally good accommodation. Beefsteaks and tea, cheese, and cider formed the usual fare. A farmer took him on the next stage of the journey, twenty miles to Digby, for £4. There he took a steam packet the thirty-six miles to Saint John, and as soon as the river was frozen solid he set out in a sleigh which he hired for £7. At one point one of his horses fell through the ice and had to be temporarily choked to be helped out of the water—an experience not uncommon in North American winter travel. Except for one inn called candidly "Poverty-hall," Head found the New Brunswick taverns good, though perhaps not well heated. His narrative, grammatically askew here and there, is vivid in its description of conditions:

I had purchased a buffalo apron, or two skins of the animal sewed together and lined with baize—an article of the greatest use and comfort: it was my friend by day and night. Of a substance warm as sheepskin and of very large dimensions, my knees and feet were defended from the weather during the many hours I was necessarily exposed in open carriages; and it supplied the insufficiency of covering in the beds and places where I lay down to rest at night.

Midway through the journey came a stretch eighty-three miles long which could be negotiated only on foot. Head walked on snowshoes while his guides pulled toboggans packed with his supplies and belongings. The winds were heavy and his feet were often swollen from the friction of the unfamiliar trappings. At Rivière du Loup he finally found a road again and paid eight guineas to be carried the final 111 miles to Quebec by carriole.

By 1830 the route was somewhat improved. There was a fair road from Fredericton to Woodstock, and for the next fifty-six miles to the Aroostook River, though some of the bridges were apt to be out. Then came nineteen miles to Grand Falls over a road that was usually unfit for use. The thirty-six miles along the Madawaska River were travelled by canoe. Then came forty-five miles across Lake Temiscouata. From

The Sleigh Club met one frosty morning in 1837 in front of the army barracks at Saint John; the scene was captured by R. G. A. Levinge, one of the officers stationed there, and lithographed the next year in London. There was no end to the variety of sleighs and carrioles in use in those days, even to tiny dog-drawn models for children.

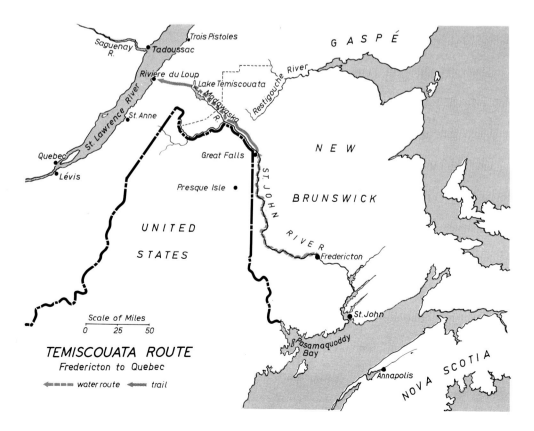

TEMISCOUATA ROUTE
Fredericton to Quebec
water route   trail

the banks of Temiscouata to Notre Dame du Portage, on the St. Lawrence, was a thirty-six-mile portage road of which hardly more than five miles was practicable for wheels. For the rest of the trip to Quebec, however, carriages could proceed at five or six miles per hour over a good road. Small wonder that even with such "improvements" this route was infrequently used! About the only regular traveller was the postman on his monthly round between Quebec and Halifax. Lieutenant E. T. Coke, who crossed the Temiscouata Portage in 1832, has left a vivid account of how surprised that lonely courier was to meet another wayfarer.

By mid-day we arrived at the river St. Francis, where we met the royal mail upon its way from Halifax. The letter bags were fastened upon a low sledge drawn by a single horse, which was moving quietly along, cropping what little grass grew by the road-side. The guard, fifty yards behind, was taking it equally leisurely, amusing himself by blowing through his tin horn and listening to the echo of the unmusical notes he produced as they resounded amongst the distant hills. The meeting was unexpected on both sides, and as he came suddenly round a turn in the forest, raising his hand to salute us, he slipped over a stone, and fell upon his back in a mass of mud and water; but rising again immediately, with the most enviable unconcern, he stood up to his knees in it, answering our numerous queries. He travelled over the road, or seventy-two miles, without meeting a human being in three months, and I will bear witness he had no sinecure. . . .

From the military point of view the Temiscouata Trail ran uncomfortably close to the United States border in the upper stretches of the St. John. In 1825 Sir James Kempt, then lieutenant-governor of Nova Scotia, planned an alternative route which had the added advantage that it would encourage the settlement of eastern New Brunswick. The Kempt Road, as it came to be called, went generally overland from Halifax to Shediac, followed the coastline to Chatham, Bathurst, and Campbellton, and then cut across the Gaspé peninsula to Métis on the St. Lawrence. It was partially open in the thirties, but even many years later it was often impassable to all but pedestrians during the rainy seasons of spring and autumn, for it was pushed through forest, swamp, and mountains with grades frequently as high as fourteen per cent. By 1849 it was apparently adequate for vehicles. That year J. F. W. Johnston travelled over it to Halifax in a light wagon. He found the one-hundred-mile section across the Gaspé a three-day trip, and rather rough and lightly used: the only vehicle he met belonged to the postman, who was on his weekly round. At Bathurst, however, he passed over the harbour on a half-mile bridge, and after that impressive sign of civilization he encountered many more settlers on the rest of the way down the coast.

In the competition between the Temiscouata and Kempt roads, the older was destined to win. Today drivers along the Trans-Canada Highway speed over the same route that Sir George Head followed so arduously 150 years ago.

4

Upper Canada

The road from York (Toronto) to Kingston was one of the most important when J. P. Cockburn painted it in 1830. Walking was still common—it was often the easiest way to travel.

IN EARLY ONTARIO we find, for the first time in Canada, a system of roads being developed almost as soon as settlement began. It was a system, moreover, which did not just grow, but was planned in its broad outlines from the start.

Apart from the Indians, and a few hundred French who were established near the Detroit River before 1759, the settlement of Upper Canada began with the arrival of the United Empire Loyalists in 1784. In their flight from the new republic the Loyalists made their way to the Canadian border by any means they could, some in boats, others in carriages or carts, but many on foot. In their new home they found little that could conceivably be called roads. Ontario was then a wilderness of forest and meadow, with an occasional fort, fur-trading post, or mission station as a sign of civilization. There were paths and portage trails, it is true, and the remains of a few short military roads which had been built near Niagara and Kingston during the Seven Years' War a quarter of a century earlier—but nothing else.

The Loyalists arrived in two main streams. One group came up the St. Lawrence and settled on its shores and that of Lake Ontario as far as the Bay of Quinte. They depended on water for transportation; it was a gruelling journey upriver in boats which had to be unloaded and dragged through each stretch of rapids, but that was easier than cutting a road. Land transportation was slow to develop in this area. As late as 1791, when in theory a road existed from Montreal to Kingston, it had in fact been opened only between Montreal and Lake St. Francis, and between Cornwall and Prescott. Both stretches paralleled rapids. A horse-boat was used to cross Lake St. Francis, and sailing ships plied Lake Ontario. Some years later a traveller remarked that even then no one ever thought of using the road all the way because water travel was so much easier.

The second Loyalist stream came across the Niagara River and up the Niagara Peninsula to the head of Lake Ontario. Quite early these settlers began to build roads where they could not travel by water. One of the first led around Niagara Falls, following the old portage trail; this was being used by carts as early as 1789. Four years before that a path had been blazed from Niagara to Ancaster, a few miles west of Hamilton.

The first need was for local communication, but after Upper Canada was created as a separate province in 1791 road-building began on a larger plan. It was obviously desirable to maintain contact between the two main centres of Loyalist population, as well as with the smaller settlements along Lake Erie and the British outpost at Detroit. For some time an "express" had been making its way each winter from Montreal to Detroit and back, carrying mail for the soldiers and merchants; but the express was nothing more than a courier, a white man with one or two Indian guides, travelling on snowshoes along the river and lakeshore. The express completed one trip a year, and for that no road was necessary—just an axe to clear away brush. Something better was wanted, particularly after Lieutenant-Governor Simcoe chose for his capital a site on the shores of Lake Ontario, partway between Niagara and Kingston, and called it Toronto.

Road-building in Upper Canada was also accelerated by the nature of its colonists. By this time Europeans had been living in North America for well over a century. The Loyalists who came to Canada had in many instances given up comfortable homes in long-settled regions of the United States. They knew roads could be built in North America, and were ready to work hard to build them. Many other settlers were soldiers, accustomed to the discipline and labour of army life. Upper Canada, moreover, could to some extent draw upon the resources of the older provinces. Thus, while its settlers depended upon the lakes and rivers as long as these routes could meet their needs, quite soon they began to develop land communications. As the population swelled with immigrants from Britain and the United States, settlement moved inland and the road network spread more rapidly. Lake Ontario remained for years the "main street" for long-distance travel, but it never dominated transportation in Upper Canada to the same extent that the St. Lawrence and Atlantic did in the east.

With the same imagination and drive that caused him to found an entirely new capital in the woods, Governor Simcoe pursued an aggressive policy in laying out roads. He conceived of two main routes, or "streets" as he called them, in imitation of the straight, enduring "streets" the Romans had built across England nearly 1,500 years earlier. One of them, Yonge Street, would run north and south between Lake Ontario and Lake Simcoe along an important trading route. The other would extend east and west, linking one end of the province with the other. Yonge Street was pushed through by 1796, the second some years later. Neither was by any stretch of the imagination a Roman road—travel along them was difficult and for a long time strictly to be avoided if possible—but their conception was on a grand scale new to the colonies.

The typical pioneer road was criss-crossed with ruts and pocked with stones. This one of 1837 passed in front of the log council house and church of a Methodist Indian mission near the mouth of the Credit River. (From a sketch by Mrs. E. Carey.)

Yonge Street

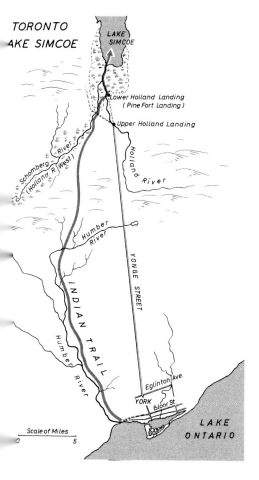

Long before the coming of the white man, Indians had worn a trail northward through the woods from the region of Toronto on Lake Ontario to Lake Simcoe. When Augustus Jones, Provincial Land Surveyor, was surveying in 1792 he came across this "Indian foot-path," and in the following year Governor Simcoe followed it during an expedition to Lake Huron. It was a difficult journey, Mrs. Simcoe recorded in her diary. At one point the party encountered "a terrible bog of liquid mud," but "the Indians with some difficulty pushed the canoe the Governor was in through it." On the way back one of the Indian guides decamped with most of the food; the remaining Indians lost their way, and it was only Simcoe's compass which led them back to Lake Ontario before their supplies ran out.

All traces of this old portage route have long been obliterated, though the remains of a corduroy road persisted for many years in the swamp at Lake Simcoe near Holland Landing. We know, however, that it ran from the mouth of the Humber River, in the west end of modern Toronto, to the eastern branch of the Holland River. In some more northerly points it followed the present course of Yonge Street. From the Humber to Upper Holland Landing—the old Indian landing place for canoes—the portage was nearly thirty miles. The Lower Landing, one and a half miles closer to Lake Simcoe, was used by boats of deeper draught; at that point the Holland was about twenty-five yards wide.

From Lake Simcoe the Severn River runs down to Georgian Bay. A land link from the new capital at Toronto to Lake Simcoe would therefore be the first step to the Upper Lakes. With an adequate road, soldiers and merchants could avoid the roundabout trip across the Niagara portage and Lake Erie. Simcoe accordingly planned such a road, which he named after Sir George Yonge, then British Minister of War. The route was partially surveyed by Augustus Jones in 1793–94, but it was blazed only in part, and for a time any improvement was left chiefly to the settlers. Towards the close of 1795, however, Simcoe ordered it opened the rest of the way to Lake Simcoe, and as soon as the New Year celebrations were ended the Queen's Rangers set to work. They blazed the thirty-three-mile road from York (as the capital had meanwhile been renamed) to Pine Fort Landing, on Lake Simcoe, in about six weeks.

It was not an easy road to travel, even on horseback. Less than two months after its opening Mrs. Simcoe rode along the southern portion and reported, "The Road is as yet very bad; there are pools of water among roots of trees and fallen logs in swampy spots, and these pools, being half frozen, are rendered still more disagreeable when the horses plunge into them." In fact, a great deal of work still had to be done to make Yonge Street an effective highway. Simcoe's route lay over a series of steep hills, whose sides were in many cases sticky, wet clay. We are told that in 1797 one of the founders of Richmond Hill, the village atop the last of these rises, found it impossible to get his wagonful of goods up the steep slopes in the normal manner: finally he had to dismantle the vehicle and drag wheels and axles and other equipment uphill by strong ropes! In 1799 the North West Company chose Yonge

Street as an alternative route to the Upper Lakes and the west, avoiding the difficult canoe trip up the Ottawa River. The company poured thousands of pounds in the next dozen years into improving the road so that it could be used by wagons loaded with boats and trade goods. Yet the solution could not have been altogether satisfactory, for in 1801 a survey reported Yonge Street as still cluttered with unburned trees and brush, and for the greater part impassable for any carriage because of logs lying on the road. In the late 1830s Yonge Street was partially macadamized, but the nature of the soil prevented the twelve miles so improved from remaining long in usable condition. When it came time to build a modern thoroughfare from Toronto to Lake Simcoe in 1951–52, surveyors did not repeat Simcoe's error. They laid out Highway 400 through easier country, 7.4 miles west of Yonge Street.

Dundas Street and the Danforth Road

The second road Simcoe planned was to run five hundred miles from Detroit to the Quebec border. He did not live to see it finished; yet by the time of his death in England in 1806 a remarkable amount had been achieved.

The first stretch of Simcoe's Dundas Street was opened, like Yonge Street, to join two navigable bodies of water. Soon after his arrival at Toronto in 1793 the governor ordered a captain and one hundred soldiers to clear the route from Dundas, then an important port at the head of Lake Ontario, to the Thames River, which drains much of southwestern Ontario before emptying into Lake St. Clair. The new road would eliminate the long detour by Queenston and Fort Erie for travellers to the west. Like Yonge Street, it too was named to political advantage, in this case after Henry Dundas, Secretary of State in the Imperial government. Many still call it "the Governor's Road."

The soldiers who opened Dundas Street presumably made use of a few roads which had already been blazed by Loyalists east and west of Delaware. A survey map of 1793 shows one road from Delaware to the Moravian Village near the modern village of Bothwell, another to the entrance of Kettle Creek on Lake Erie, and a third to the Mohawk Village close to Brantford. All were well-known Indian trails. By the spring of 1794, with this help, Dundas Street was extended from Lake Ontario to the future site of London and, in some sections, further westward, though beyond the forks of La Tranche (Thames) the river was the most common route.

The eastern portion from Dundas to Kingston was built more slowly. Between these points Lake Ontario provided an easy passage by boat through most of the year. Bridle-paths were opened in some sections during the 1790s, but in other parts the lake shore was the only road, and over it some travellers, women among them, are said to have walked from Kingston to York. We are not told how they negotiated the river-mouths.

In 1798 the government awarded a contract to blaze a proper road from Kingston westward along the front of Lake Ontario at a price of $90 per mile. The contractor was Asa Danforth, an American who was one of this country's great early road builders. The work took three years, and would have taken longer if he had encountered primeval forest all the way. Fortunately, the stretch from Kingston to Trenton passed through a region already settled by Loyalists, and local roadways had merely to be improved and joined. Danforth set up headquarters at the Finkle Tavern at Bath, and from there extended the road to Adolphustown and (with a ferry link) to Picton, Wellington, Consecon, and the Carrying Place.

Danforth's name came eventually to be associated with the much more difficult section between Trenton and York. Work began at the western end late in the summer of 1799. The previous April Danforth had signed an agreement on standards. The highway was to be cleared thirty-three feet wide throughout its length; of this sixteen and a half feet (as far as possible in the middle) were to be cut smooth and even with the ground. Bridges and causeways were to be built sixteen and a half feet wide wherever necessary, and high enough not to be washed away. Slopes were to be safe, gradual, and easy, and wide enough for a sleigh or carriage to pass. By early December sixty-three miles were opened for safe travel by sleighs and wagons between York and Smith's Creek (Port Hope). At that point Senior Surveyor W. Chewett inspected the work for the government. His report shows just what "opening a road" meant in 1799 and for many years afterwards.

Above, P. J. Bainbrigge: "Bush Farm near Chatham." Right, J. P. Cockburn: "Kingston."

The inspector recognized the great difficulties in blazing a way through a wilderness in which there were only four settlers near the road in more than sixty miles. He found that some hollows needed filling, some "knowls" and hillocks levelling, and that rotten logs and stumps, as well as underbrush, should be removed. The bridges and causeways were generally good, but an exception was the bridge over Barber's Creek, "which shakes, or trembles" when a horse or carriage was upon it. Certain other bridges were not long enough, and cattle or carriages would be mired in attempting to negotiate them; but this deficiency, he said, could be easily remedied in the spring. The hills were sometimes too steep. That on the east side of the River Nen (Rouge)— the most difficult part of the route—was unsatisfactory, although the contractor had taken all pains in his power. "I saw a loaded Ox Sled go down with ease," said the inspector, but it was helped by snow sixteen to eighteen inches deep; he was sure carriages would have to unload to get up the hill, and if it were icy nothing could go either up or down—though chains on sled-runners might help, or vehicles might be hoisted or lowered by block and tackle.

All in all, Chewett considered the completed section a "good" winter road over which oxen might haul a loaded sled sixteen or eighteen miles within the daylight of a winter day, or horses perhaps thirty-five to forty miles. A summer road suitable for carriages would be impossible, however, until there were enough settlers along the road to keep it in constant repair and clear of brush and logs. Brush would grow up at every stump, Chewett pointed out, and he himself had noticed that between the time work parties made their way eastward and then started back to York, some forty or fifty trees had fallen and had to be hauled out of the way. He recommended therefore that the land along the road be turned over in two-hundred-acre lots to people prepared to settle it immediately and keep the route in good condition. He further recommended that none of the roadside property be set aside as Crown or Clergy Reserves: as the law stood at that time one-seventh of the land in each township had to be reserved for the government, and one-seventh for the support of the Protestant clergy, but these lots could be located anywhere in the township, and since they were rarely occupied their presence along the road would create gaps in the do-it-yourself system upon which maintenance depended.

Chewett felt that Danforth should be allowed half the amount due him so that he could pay his workmen and obtain credit for supplies. Otherwise he would be ruined and the aims of the government frustrated. He felt that if Danforth were also given an extra month to complete the work, all necessary improvements could be effected in the spring. But, he noted cautiously, his survey should have been made earlier in the year, "for when an inspector is almost frozen he cannot act as he ought to." The contractor could have asked for the inspection in January, and if he had done so, Chewett pointed out, everything would no doubt have seemed satisfactory under three to seven feet of snow. As for what spring floods might do to the newly blazed road and its bridges he could not tell, for he had never seen the high-water marks of rivers and creeks.

Clearly Danforth had undertaken a project that, however well carried out, was unlikely to remain long in satisfactory condition. Frosts and spring floods were certain to wreak havoc, and along the entire route there were very few settlers and many reserved lots and uninhabited regions where it was nobody's business to do even the most necessary repair or maintenance work. In 1802 Chewett and two other inspectors were appointed by the Surveyor-General to report on the 103-mile section to the River Trent. They found general fault with its width, bridges, causeways, and numerous other matters, including the stumps still standing in its course. Their detailed estimate of the expenditure necessary to put the roadway "in the state it should have been in on the day of delivery to make it conformable to the contract," amounted to £2100 5s. In spite of their harsh judgment our sympathies are largely with Asa Danforth.

By 1800 Danforth had opened his road westward beyond York as

far as Ancaster—a difficult piece of work because of the great ravines made by the Credit and other rivers on their way to Lake Ontario. Nine years after the creation of Upper Canada roads had been blazed at one time or another from the border of Lower Canada to the head of Lake Ontario, and beyond to the western end of Lake Erie. But they quickly fell into disuse when water transport was easier, and the greater part of the entire route was for a long time traversed only by Indian trails and blazed paths. Crown or Clergy reserves, which often hindered or blocked other roads and settlements, were not an issue here; nevertheless the route could not have been negotiated by carriage or wagon, except over short distances, and though more primitive travel on foot or horseback was possible, it was extremely rigorous and not often attempted.

There was no demand at the time for stage travel, and it could not have been met if there had been, though in winter it was usually possible to travel anywhere, road or no road. Danforth blazed his highway through the province as well as anyone could in the circumstances, but it speedily reverted to a wild and impassable state except where villagers or groups of farmers were interested in keeping it open. During the navigation season west-bound mail was always carried by boat to York and Niagara, and travellers too preferred to take to the water. In winter the post was carried by a white courier from Montreal to Kingston, and from there by an Indian courier who was expected to continue through York to Niagara. It took sixteen to eighteen days for a letter from Quebec to reach York, and the service was provided only once a month. Couriers also carried the mail to western Ontario, but as late as 1807 three months elapsed between every trip to the Windsor area.

In the winter of 1816–17 a new route—the Kingston Road—was opened. It followed Danforth's blaze in some parts; in others it ran closer to the lake. For many years it too was difficult to use except in winter. When conditions did improve, the mails were carried along it by stagecoach, but this was usually impossible in the spring, and as late as 1831 horseback couriers were still required. At that time one mail bag was sufficient for all the western settlements.

A relic of the Danforth Road still stands northeast of Cobourg—one of the oldest inns in Ontario, once the Half-Way House between Kingston and York. The tradition persists that Sir Isaac Brock stopped there overnight during the War of 1812 on his way to Niagara and death. The building is a large three-storey structure built with the massive beams and wide boards characteristic of the times. Many of the planks are from fifteen to eighteen inches wide by two inches thick and were evidently cut laboriously by whip-saw. The ten-by-twelve-inch beams were hand hewn by broadaxe. On the slope of a hill to the rear are the remains of a cider mill that provided some of the tavern's refreshments. Other fare was prepared in the six-foot fireplace and inside bake-oven. The western portion of the third floor was the ball-room, and both its size and the solidity of its construction may be gauged from the fact that today it still sometimes accommodates 150 people for square dances. No more interesting inn has survived for more than a century and a half.

Other Roads

While the government concentrated most of its meagre resources on these two trunk routes, many lesser roads had to be left for the settlers themselves to construct. New stretches were opened, or old ones improved, in a rather piecemeal fashion as the occasion allowed. Each settler was required by law to clear a roadway along the concession on which his lot fronted. One means of building short stretches through thinly settled areas was by subscription—that is, by the public soliciting of funds.

This technique could be remarkably successful. On the Niagara Peninsula, for example, the bogs of the Black Swamp sprawled across nearly two miles of the route to the head of Lake Ontario, threatening and sometimes totally barring the passage of the traveller. The obvious solution was to build a causeway across the swamp, a job which it was predicted would cost thousands of dollars and months of labour. The government was unable to tackle it, but this did not stop the men who used the road. At taverns and mills along the route they posted subscription papers where, as they said, "those gentlemen who are inclined to aid so necessary a work will have an opportunity of adding their donations." Work began almost immediately, and the task proved far less formidable than it had appeared. After five weeks the causeway was almost two-thirds complete. Only 240 rods (three quarters of a mile) remained to be finished at an estimated cost of $1.50 per rod. For speed of construction this was an Upper Canadian record, its sponsors boasted in a newspaper announcement. "Three hundred and sixty dollars more, gentlemen," they added, "and the work will be completed." It must be remembered, however, that causeways of those days (or "crossways" as engineers sometimes called them) were formed mainly of stumps and logs. Under the force of frost and spring floods they usually degenerated into what Mrs. Simcoe described as "that terrible kind of road where the horses' feet are entangled among the logs and water and swamp."

In 1859, Emerson Taylor's tavern on Dundas Street was the halfway house for stagecoaches running between Hamilton and Toronto. It stood at Springfield-on-the-Credit (now Erindale), once the site of the old Indian mission shown on page 40.

As already mentioned, several sections of Dundas Street were also opened by private initiative before the government road-builders arrived. The early settlers of Oxford County were notably ambitious in this regard. They were well aware of the importance of adequate roads in raising the value of property, and despite the difficulties of pioneer life they set to work on the roads almost immediately after their arrival in the district. In a single year they managed to cut and bridge a road from Burford (a few miles west of Brantford) to the Thames through a wilderness of twenty-five to thirty miles. According to a contemporary account the cost was met by Thomas Ingersoll, whose son Charles founded the town of Ingersoll which lies along this route. A subscription was taken to extend the road beyond London, but the funds so raised were insufficient, with the result that for some time the way was left unfinished, and was passable only by sleighs in winter.

The final link in the trans-provincial road system was built by another private landowner rather than the government. Colonel Thomas Talbot was one of the most colourful and controversial of Ontario pioneers. He had founded a settlement on Lake Erie in 1803 and was so successful—or had so much influence in high places—that eight years later control of most of the London and Western Districts was turned over by the province to him. While he did not own all the territory, its development was left in his hands; he alone decided what persons should be allowed to take up land and where they should live. He was a hard-drinking autocrat, disliked intensely by most of the settlers for the way in which he exercised these semi-feudal rights. The fact remains, however, that his domain quickly became one of the most flourishing areas of Upper Canada, with some fifty thousand inhabitants, according to Talbot, by 1837.

Before he could claim title to his land, every Talbot settler was expected to build a log cabin, clear five acres, and open a roadway. Talbot also ordered the construction of highways, beginning in 1809 under the direction of Colonel Mahlon Burwell. The first sections were completed in 1811 under considerable difficulties. Bad weather held up the operations, and supplies were almost impossible to obtain for man or beast: the road-builders had to buy their wheat as it stood in the fields, thresh it themselves, and carry it to the nearest mill where they also had to do the milling. Talbot's highway system was gradually

extended, however, and eventually two main branches were completed. One ran across southwestern Ontario from Fort Erie to Sandwich (now part of Windsor), where it joined a government-built road to Sarnia. The other ran from Port Talbot, on Lake Erie, through Port Stanley and St. Thomas (also named after Talbot, though he was assuredly no saint!) and on to London, joining there with Dundas Street. Undoubtedly one of Talbot's concerns in building these roads was to enhance the value and extent of his own grant which increased with every new settler, and to accelerate its settlement at the expense of London and the district to the north. Talbot Street soon became known as the best road in Upper Canada. A stage service was operating along its length for a time in the 1820s, and it was later one of the first roads to be improved from the original corduroy by more advanced types of construction.

Throughout the early decades of Upper Canada, new roads were the keys which opened to settlement the fertile inland areas of the province.

In the smaller towns, streets long followed the most convenient route around stumps. This is Sixth Street, Chatham, in 1838, painted by P. J. Bainbrigge.

Pioneers in Ontario were not restricted to a narrow river valley as the early colonists in Quebec had been. But while there were thousands of square miles of good land north of Lake Ontario and Lake Erie, these were rarely connected by navigable waterways to the St. Lawrence system. The only way to carry supplies to this frontier, and to carry farm produce from it, was by land. At first this meant travelling on foot. As soon as possible crude roads were cleared—from Dundas to Waterloo County in 1799–1800, for example, to serve the needs of German settlers who were just beginning to enter that district. That particular road passed through the notorious Beverley Swamp which for years provided the raw material for travellers' tales of terrifyingly narrow escapes.

The Ottawa district was opened to settlement in 1816. Two years later a blazed trail, known rather grandly as the Richmond Road, was cleared by soldier settlers from Perth to Richmond. Another was pushed northward from Brockville. On neither was it possible to travel at a speed of much more than two miles an hour. G. J. Mountain,

By 1835 King Street, then Toronto's main thoroughfare, was broad and adequately surfaced—in dry weather. Board sidewalks protected long skirts from dirt. This painting by John Howard looks eastward from York Street towards Yonge.

later to become third Anglican bishop of Quebec, followed the Richmond Road to Perth and said it could be negotiated only in three stages of seven miles each; each stage took three hours, and though he had been over some dreadful roads elsewhere, they were "turnpike and bowling green" compared to this one. We have an even more graphic description from John M'Donald, who travelled from Brockville to New Lanark and New Perth with a number of immigrants in 1821. Their wagons were frequently upset, and in these accidents one boy was killed, a man had his arm broken, and several other persons were hurt. M'Donald's own driver, despite his care and skill, blundered into a mire so tenacious that the horse could not extricate itself even after

it had been unhitched. Finally the men had to use handspikes to free the hoofs of the animal from the sticky clay. The wagon was pulled out by a passing farmer with a team of oxen, and after that when the party came upon bad patches they pulled up the roadside fences of earlier settlers, laid the poles down to form a temporary corduroy, and after passing the mire, rebuilt them. Their courtesy was exceptional, for in many other instances rail fences were treated as fair game, and were torn down with no thought of reconstruction.

Normally roads appeared shortly after the first settlers. In 1827, however, the carriage-width Huron Road was blazed by the Canada Company before anyone had arrived, to ensure prospective pioneers at least primitive land communication between Lake Ontario and Lake Huron. The road ran from Guelph to Goderich, largely through land the company had acquired and was anxious to sell. Other roads opened as the pressure for settlement continued; among them were the Huron-tario Road from Port Credit on Lake Ontario to Collingwood on Georgian Bay, and the Garafraxa Road, roughly paralleling it to the west, from Oakville to Owen Sound. By the early forties there were about six thousand miles of post-road in Canada West.

Many more were needed, however. Petitions flowed into the legislature from settlers in the rear townships seeking improved communica-

George Cockburn's Hotel at Baltimore, near Cobourg, in 1878. Cockburn also bred and imported horses. The tavern, now a residence, still stands.

tion with "the front." A typical petition in the Public Archives of Canada was signed by fifty-one inhabitants of the Fenelon Falls district in 1842, calling for a road to be completed from Lindsay to Windsor (now Whitby). A line of communication with Lake Ontario, they said, had been promised by the government before they agreed to settle the area. Yet at the time of petition "the lamentably imperfect roads which have been cut through the wilderness by the needy and thinly scattered pioneers of the forest are such as can be traversed for only a short portion of the year," and they were without access to a market for the remaining months. Until a new line of communication was developed, in fact, settlers in the entire "back lakes" (Kawartha) region were restricted to travel by canoe, ox-cart, and sled; in many regions the most feasible method of carrying supplies was still to pack them in on a man's back.

Even the best parts of this early road system were none too good by today's standards. Travellers and settlers lumped the trunk routes with all the others in Upper Canada as deplorable. Except perhaps for a few weeks in midsummer when the roadbeds were dry, and midwinter when they were frozen and covered with snow, few routes could be followed except on foot or horseback. In a letter to the *Niagara Herald* in 1801, one man complained that his own misfortunes on a journey had been bad enough, but he had seen others worse off—families whose wagons had broken down, fallen into deep gullies, or been trapped in bridgeless creeks. A quarter of a century later two English travellers, Captain Basil Hall and his wife, found conditions not much improved. They were in a "stage wagon" on the Kingston Road in midsummer when they encountered corduroy interspersed with mud holes so deep that the wagon wheels were swallowed by them. The trip was a succession of jolting over the logs and plunging into the mire; the Halls had never met anything like it in their lives, but to their credit they did their best to laugh about it. Such roads, observed one self-styled lady, made Upper Canada "a vile country except for low people"; and she added, "Why don't they have the roads Macadamised, I should like to know?"

Travellers and guidebooks, indeed, were unanimous in condemning early Ontario roads. *The Emigrant's Guide to Upper Canada*, published in 1820, described them as few and poor, with rough surfaces, wretched bridges, and inadequate service at the inns; but it was admitted that they were still "moderately commensurate with the retarded progress" of the country. Roads, said another critic, were the worst feature of Upper Canada, even though the settlers in some areas were making great improvements.

Once the traveller left the main route his situation was indeed melancholy. In 1837 Anna Jameson journeyed to Chatham. Her vehicle was a farmer's cart, and for much of the way she followed Talbot Street in relative comfort. Then the driver turned off to a side road—and trouble began. Mosquitoes were thick in the forest, but they were only part of the discomfort:

The road was scarcely passable; there were no longer cheerful farms and clearings, but the dark pine forest and the rank swamp, crossed by those terrific

corduroy paths, (my bones ache at the mere recollection!), and deep holes and pools of rotted vegetable matter mixed with water, black, bottomless sloughs of despond! The very horses paused on the brink of some of these mud-gulfs, and trembled ere they made the plunge downwards. I set my teeth, screwed myself to my seat, and commended myself to Heaven—but I was well-nigh dislocated!

At length I abandoned my seat altogether, and made an attempt to recline on the straw at the bottom of the cart, disposing my cloaks, carpet-bags, and pillow so as to afford some support—but all in vain; myself and all my well-contrived edifice of comfort were pitched hither and thither, and I expected at every moment to be thrown over headlong.

When that stretch of road finally ended, Mrs. Jameson rested for a brief time at a crude log inn before setting out again. This time there was no road at all through the forest—simply a blazed path over damp black earth into which the cartwheels sank a foot deep. Occasionally the travellers encountered stumps and roots over which the cart had to be lifted or dragged; sometimes they traversed swamps which had been crudely filled in. It was a great relief, she concluded, to emerge into the Thames valley; and the sight of Chatham "made my sinking spirits bound like the sight of a friend."

Charles Dickens visited Canada in 1842. It is unlikely that he left the main roads, yet this is what he saw:

There was the swamp, the bush, and the perpetual chorus of frogs, the rank unseemly growth, the unwholesome steaming earth. Here and there, and frequently too, we encountered a solitary broken down wagon, full of some new settler's goods. It was a pitiful sight to see one of these vehicles deep in the mire; the axletree broken; the wheel lying idly by its side; the man gone miles away to look for assistance; the woman seated among their wandering household goods with a baby at her breast, a picture of forlorn, dejected patience; the team of oxen crouching down mournfully in the mud, and breathing forth such clouds of vapour from their mouths and nostrils that all the damp mist and fog around seemed to have come direct from them.

But under some circumstances a traveller might be delighted to see

One year before Confederation, the steamship was still the most comfortable means of travel between Picton and other points on Lake Ontario; but the wide variety of wheeled vehicles attests to the use made of roads. ("Picton Harbour," engraved from a sketch by G. Ackerman.)

even a corduroy road. About the same time that Dickens was in Canada, Sir Charles Lyell, one of the most famous geologists of the day, became lost in the woods north of Toronto. With him was Thomas Roy, who two years earlier had surveyed that very area for new settlement. After a day and most of a night on horseback, they were picking their way through a bog by moonlight when they saw the outlines of a corduroy strip. With it as a guide they quickly located their inn. "I shall always, in future, regard a corduroy road with respect, as marking a great step in the march of civilization," Sir Charles remarked.

Under such conditions makeshift means were needed to provide even irregular mail service to the back settlements. In later years stagecoaches served the main roads, but beyond them letters were carried at best by rough wagon or horseback, and often by foot. Frequently settlers clubbed together to have their mail brought to them from the nearest post office on "the front"; otherwise it waited there until called for.

It was unavoidable that the roads should be bad. As we shall see, the technology to improve them was not lacking. Macadamized roads were being laid in England, and experiments with plank roads were under way in Canada. The citizens were well aware of what improvements could mean. One newspaper editor proposed that a million pounds, no small sum in those days, should be borrowed to relieve this "crying and degrading evil" which was holding up the economic advance of the entire province: he predicted that the resulting increase in trade and communication would put money in everyone's pocket, and pay for itself twenty times over. Yet the money was not to be had. There was far less currency in circulation at that time than today, even allowing for the difference in population; to the average Canadian a silver dollar seemed as large as a cartwheel. Nor did the governments then enter into deficit financing to carry out public works. In Upper Canada it was the same story as elsewhere—not enough men, money, or time to serve a great area that was developing rapidly. Soldiers were used occasionally to build roads, and when funds were available contractors were hired; but to a large extent the government's role was to encourage settlers to build and maintain their own roads, to enforce a system of statute labour which would ensure that the essential minimum of work was performed, and to direct these amateur road-builders.

Provision for statute labour was made in some of the earliest legislation. The Highways Act of 1793 compelled all inhabitants to work on roads and bridges, but each man for not more than twelve days a year. Five years later this was amended to make the time somewhat proportional to the assessed value of property owned, as follows:

Assessment	Days
Up to £25	2
£25 to £50	3
£50 to £75	4
£75 to £100	5
£100 to £150	6
£150 to £200	7
£200 to £250	8
£250 to £300	9
£300 to £350	10
£350 to £400	11
£400 to £500	12

Additions, but not *pro rata*, were made for higher assessments, up to £3500, an extra day being added for each additional £100, £200, £300, and £500 in four progressions. It is apparent that the rich landowner was favoured. This inequity stayed, however, and another remained until 1840, when a provision was effected requiring two days' labour of those not listed on assessment rolls. A team and wagon were considered equal to two days' labour by a man.

The road-builders were not completely without money. There were fixed rates of commutation, under which those who wished and could afford to could pay others to do their work. This commutation was compulsory for town-dwellers and landowners liable to more than six days' work, and after 1840 for those living within half a mile of a macadamized road. The payments so obtained provided some funds for road maintenance, and so did the fines imposed on those who failed to do their statute labour on time. These sources of revenue might be supplemented by a levy of up to £50 (later increased to £100) which could be ordered on a district by the justice of the peace, and there was a road tax of twenty shillings per year on absentee owners of more than two hundred acres. Far more important, eventually, was the annual road grant made by the legislature. This rose from £1,000 in 1804 to £6,000 in 1812, and to £20,000 in 1815, the needs of wartime transportation no doubt being responsible for much of the

later increase. By the late 1830s the sum was still larger, totalling £100,000 in the period 1836–40, again partly owing to military requirements in a period of unrest and rebellion. Despite the increased amounts, there was no diminution in the complaints and petitions with regard to roads; the mere reading of them consumed an enormous amount of time in the Legislative Assembly.

Around 1840 several stretches of macadamized road were built and paid for by turnpike trusts, which charged a toll for the use of these improved thoroughfares. Each turnpike was administered by a board of commissioners who were not paid, who could not themselves be involved in any road contract, but who were responsible for the appointment of engineers, surveyors, toll collectors, and other personnel; for the number of tollgates (which often were let to the highest bidder); and for the tolls charged. The tolls varied from place to place, but not greatly. There were many complaints, however, that the roads provided under this system were poor.

With the Act of Union of 1841, which brought Upper and Lower Canada together under a single legislature, control of all but the most important roads was transferred from the province to a number of district councils. In 1849, when municipal government was organized in Canada West (Ontario), elected county and township councils were made responsible for the roads. At the same time an act was passed authorizing joint stock companies to build roads and bridges and to charge for their use. Perhaps in general these companies were no more popular than that which operated the Cobourg–Port Hope turnpike, which had its tollhouse burned in protest against high rates and low standards. Municipalities were, however, permitted to buy shares in such companies and eventually take them over. During the 1850s further legislation made it possible for them to borrow money for such purposes; the money was usually obtained in England, with financial responsibility assumed by the province. A large amount of municipal borrowing was lost in railway speculation during the mania that swept Ontario in the fifties and sixties, but a certain amount was spent on roads—and that was in no sense wasted.

Making a corduroy road (from a contemporary sketch).

5

Early Road Construction

NO MATTER where pioneer families settled in early Canada, they had to find some way of maintaining contact with the outside world. Without it they would be no better off than Robinson Crusoe stranded on his desert island. They would be able to support some sort of makeshift existence with what was available in their immediate area, but there would be no means of keeping in touch with family and friends back home, getting their produce to market, or obtaining pots and pans, nails, warm wool clothing, or other manufactured essentials.

For a few months each winter, travel between settlements was relatively easy. Snow filled the hollows and flattened the bumps of the rough ground, and ice bridged even the widest streams and all but the largest lakes, though not without many hazards for travellers. Since during this period there was little to be done on the farm, the early settlers took advantage of their good fortune. Nearly every family had a team of horses and a home-made sledge—often no more than a wooden box on runners, but adequate for the purpose. Bundled up in buffalo robes and bearskins, families often travelled many hundreds of miles to visit friends and relatives in other parts of the country. Particularly in Upper Canada, where water transportation was less convenient than elsewhere, the season of ice and snow was used to carry produce to market; as late as 1825 it was estimated that at least two-thirds of the crops in that province were still being transported during the winter.

At other seasons travellers might be forced to take to the water, but the average farmer—even though he was a pioneer—was used to the land after all, and was far more likely to own a horse and cart than a good boat. In some areas it was possible to make use of natural roadways, cleared by the action of water along the shores of the ocean and lakes. Elsewhere Indian paths and portage trails were turned into crude bridle paths, over which a horse might pick its way cautiously among the roots and stumps. Gradually such paths might be widened and improved and form the basis of a proper road.

The procedure in opening a road was always much the same. If it was a new route an explorer went ahead to mark out its rough course; he was closely followed by two surveyors with compasses. Blazers then notched trees on either side to show the boundaries, and axemen chopped down those which lay in the roadway. Gangs of men followed with oxen to clear away the trunks and brush, and wagons with provisions brought up the rear. There were of course variations in this procedure. When, for example, the Canada Company blazed the Huron Road from Guelph to Goderich in 1827, the road gang cut down trees along the entire course, and then worked back over the whole route removing the debris. Usually the larger stumps were left in the ground to rot, often for years, and consequently few routes ran straight over even short sections. There were no mechanical aids. Ploughs sometimes were used instead of shovels to level hills, but in general hand tools predominated. It was hard, exhausting work. Fever and ague frequently struck, decimating the work crews, and deaths were not uncommon.

A particularly difficult piece of road-building was pushed through in 1814 in Upper Canada as part of the war effort against the United

States. A dockyard was being established on Georgian Bay at Penetanguishene, and a land link with the new base was essential, not only to carry supplies in winter when the Great Lakes could not be used, but also to guarantee a means of communication safe from enemy vessels. It was decided to extend Yonge Street northward through thirty miles of wilderness, and in the late fall two companies of soldiers were ordered to do the work under the direction of William ("Tiger") Dunlop, a young British army surgeon who had volunteered for this special service between campaigns. (Later he would become a legend as an official of the Canada Company and a chronicler of the Upper Canadian backwoods.) Lake Simcoe, their starting point, was already frozen. Just as they were about to cross it, however, there was a noise like thunder and the ice cracked all over the lake. But Dunlop refused to delay

In winter nature provided her own roads—the frozen waterways, in this instance the Kennebecasis River of New Brunswick.

the crossing. The men set out, each holding onto a rope lifeline with one hand and dragging behind him a small sleigh loaded with his knapsack, provisions, and tools. There were many accidents, none of them serious. If a man stepped on a break in the ice and fell into the cold water he was speedily dragged out by his comrades amidst, Dunlop assures us, shouts of laughter. It could hardly have been as humorous to the unfortunate victim! The trip across Lake Simcoe took six hours. At the end a huge pile of logs was lit, a space was swept clear of snow, and all sat down to a late dinner.

The road had already been started by neighbouring militia when Dunlop and his soldiers arrived on the scene. The next few miles were therefore a straight march—through three feet of snow.

Six or seven men led on snow-shoes in Indian file [Dunlop reported later], taking care to tread down the snow equally; then followed the column, also in Indian file. At about every thirty yards the leader of the column stepped aside and, letting the rest pass him, fell into the rear. By this means, after the fatigue

of first breaking the snow, he could march on a beaten path, and thus, alternating labour and rest, the thing was comparatively easy. By sunset we had made about five miles beyond the militia camp, and it was counted, considering the road, a very fair day's journey.

Once shanties had been erected, the regular troops set to work on the road. Snow, often six feet deep, prevented the use of horses or oxen, so that all provisions and supplies had to be packed by men. About half the detachment was continually involved in this work. Their comrades suffered equally from lack of draft animals, for thirty men could not haul logs with drag-ropes as easily as a yoke of oxen might have done. But there is no limit to wartime expenditure, nor to the suffering soldiers may be asked to endure. In building bridges, officers and men alike dragged logs through four feet of snow to the river's edge, then stood for hours up to the waist in ice-cold water to fix them in place. Dunlop estimated that the cost to the government was more than double what it might have been at another time of year, but happily there were no cases of accident or sickness "except casualities such as cutting feet (a very common accident even among experienced choppers), and bruises from falling trees." Just as the road was finished, and the men were looking forward to a pleasant summer at Penetanguishene, word was received that the war was over, and all were ordered to return to York and their respective regiments.

For the traveller over a newly blazed road anywhere in British North America, the way was not easy. It might be months, or even years, before the route was fully cleared, and in the meantime he had to wind in and out around stumps, avoiding pitfalls as best he could. The ground was apt to be soft and wet: the root masses of the virgin forest retained water, and swamps and bogs abounded. Dozens of streams would have to be crossed. Some might be hurdled and others would have been bridged, but it would probably be years before a man could walk any distance in dry boots.

The first improvements made use of the almost unlimited supply of wood. Thousands of trees had to be cut down in the course of building a road. Commonly they were burned or left to rot. Laid down instead across the road, at right angles to its path, the trunks provided a firm surface over wet spots—but a notoriously uncomfortable one. The logs used might be anywhere from nine inches to two feet across, and rarely was any effort made to flatten them. Often they were simply hauled into rough position some distance from their neighbours. The gaps between could be large enough to trap a horse's hoof, or even a wagon wheel, and if the space was filled with dirt or gravel it soon opened again under the hammering of the traffic. In wet weather the whole rough surface might be afloat in muck or clay. This was the infamous corduroy road, named for its resemblance to the hard-wearing, ridged cloth. Its only merit was that a cart *could* get across the swamp on it— but if the driver went too quickly his vehicle threatened to fall apart from the terrible jolting, or his horses' hoofs might be trapped between timbers.

In some districts care was taken to make as good a road as possible.

Early culverts were crude but effective. The great logs used to bridge this stream were covered with earth to make an unusually smooth crossing. (Sketch by T. H. Ware.)

The best results were obtained by splitting the logs in half and placing the flat side downwards; and if trouble was taken to place the largest logs in the softest ground, a fairly passable road could be built. Even then, however, the logs would rot, if indeed they were not first heaved out of position by frost, and if nothing further was done the roadway degenerated into a series of deep mudholes between bone-shaking stretches of still solid tree trunks.

> Half a log, half a log,
>   Half a log onward,
> Shaken and out of breath,
>   Rode we and wondered.
> Ours not to reason why,
>   Ours but to clutch and cry
> While onward we thundered.

That verse by Carrie Munson Hoople pretty well sums them up.

The most common early bridges were also of corduroy—rough logs laid over lengthwise supports. They were as uncomfortable to cross as the roads, and considerably more dangerous, for sometimes a rotten log gave way under an animal's weight. Mrs. Simcoe wrote at one point that "The horse I am now riding had once a fall through an old bridge. He now goes very carefully." At least one other traveller found a corduroy bridge strangely useful: his horse had got away while he was giving it a rest, and he was able to overtake it on a bridge where the animal was afraid to break into a run! The bridges were so lightly constructed, without engineering skill, that many travellers were amazed they could support the weights they did.

Road conditions appalled European visitors, but in truth little more could have been done in the early days of the country. Everything seemed to conspire against the road-builder. Thick virgin forest stood in his way in most areas. Countless lakes and watercourses required detours or bridging. Where the land was flat and low there were swamps which had to be filled. Elsewhere there were hills, and even mountains, with slopes too steep for easy passage. The road-builder's difficulties were accentuated by the system of road allowances, which followed the concession fronts in a straight line through every difficulty—river, lake, swamp, or valley; and still worse were the unused blocks of land reserved for clergy and Crown, at the boundaries of which the road was frequently forced to stop abruptly. Even when a road was finally opened the severe winter frosts and spring thaws did so much damage that there was no guarantee the surface would last a single season.

A typical stretch of corduroy road near Orillia, Ontario, in 1844. Horses were usually led over the logs, which were often half-afloat in swampy country. (Sketch by T. H. Ware.)

Assuredly, given enough men, money, and equipment, more could have been done. That has been proved in recent years in the far north, where roads have been built under much more difficult climatic and topographic conditions. But none of these commodities was readily available to the early Canadians. As we have seen, troops were called out on occasion to build new roads; volunteer workers and public subscription supplemented government efforts; as funds became available, private contractors like Asa Danforth were increasingly employed; but right up to the present century (and well into it, in some areas), great reliance continued to be placed on statute labour with all its inefficiency.

A traveller of 1830 waits impatiently for the ferry-man to row him across the mouth of the Trent River in Upper Canada. The scene was painted by a British army engineer, Captain Thomas Burrowes. Seven years later, when Charles Gifford passed the same way, he found "an excellent Bridge across the river, I should suppose about 700 yds long and roofed."

Nor has Canada ever fully escaped from the underlying problem of building highways over an immense and thinly populated territory.

To all this must be added an apparent pioneer readiness to accept the makeshift, to follow custom and do things the cheapest way. After a journey along the St. Lawrence River in 1833 one traveller wrote of the general apathy toward road conditions.

> Our road was now every mile getting worse, and the wooden bridges across brooks and ravines appeared to my unpractised eye to be almost impassable. My fellow-travellers, however, (an amicable young lady included) testified neither surprise nor alarm, and, of course, it did not become me to complain. The planks of the bridges were frequently so loose, so rotten, and so crazy, that I am yet at a loss to conjecture how our bulky machine and the four high-mettled steeds escaped without falling through. A sufficient supply of stone for repairs lay along the roadside, generally, too, in heaps, as gathered from the land; while timber for the bridges was certainly not far to seek. The period of annual repair had not, however, yet come round, and even then no metal [that is to say, broken stone] would be applied; the road would merely receive a sort of levelling, often, as I was assured, with the plough, and the mud-holes be in some temporary way filled up.

Liquor flowed in every road camp—whisky in Upper Canada, where there is more than one report of a pathmaster and all his crew carousing by the roadside, and rum in the Maritimes. Nova Scotia has a stream called More-Rum Creek, so named because when the road-builders reached that point the rum ran out, and they refused to proceed until supplies were refreshed.

Another serious impediment to good roads was the intrusion of politics, particularly on the local level. No trained civil service existed to originate and supervise work. Rather it was generally held that "practical" men could turn their hands to any project successfully without benefit of expert advice. Except in the case of a few trunk routes there was no apparent overall planning. For the most part each separate road, and even each stretch of road, was considered to be merely for the use of a few settlers in hauling their produce to market, and therefore of concern only to the local government, an institution introduced by early settlers from the United States. Road commissioners were politically appointed and rarely had knowledge or experience in road-building: they were more likely to be storekeepers, millers, tavernkeepers,

On unimproved back roads, a day's jaunt in the country could turn unexpectedly into a hair-raising experience. This engraving of 1871 purports to show a Natural History Society's field day.

and tradesmen with friends in high places, and often they had an axe of their own to grind in opposition to the public interest; yet with all the arrogance of ignorance they proceeded to direct operations, frequently with foremen who were equally incompetent. The local projects were so small that it would have been uneconomical to hire capable supervisors even if there had been a desire to do so, with the result that work proceeded by makeshift and employed whatever techniques were customary in the district. Farmhouses often encroached on the road allowance, and at least one barn is known to have been built in the centre of a highway. The provincial government, which footed most of the bill, did not even appoint inspectors, or demand that a satisfactory standard be achieved before grants were made. Up-to-date techniques of road construction and maintenance were available in early Canada—thanks to the work of the great British road-builders, Thomas Telford and J. L. McAdam, and the writings and practice of Canadian civil engineers like Thomas Roy, T. C. Keefer, and William Kingsford—but they were almost entirely ignored.

In any event, little could be done beyond building corduroy roads until the district on either side of the roadway had been cleared of trees and drained for cultivation. Once that was done the logs were often left high and dry, but frequently they were allowed to remain where they were because it would take too much work to move them; they were simply covered with earth, and as they rotted away more earth was shovelled on them to fill the holes.

The first real improvement over corduroy was the "common" road. It was dirt, without an artificial base, but at least it was properly drained and bridged, and the steeper hills were reduced. Ditches were dug along the side, and the centre was built up. Crude mechanization made its appearance in the construction of such roads. In 1826, when the corduroy was being taken up along Talbot Street and the road improved by a turnpike trust, a traveller saw four men and two yoke of oxen at work. One team was ploughing and the other was scraping, and together, he estimated, they were doing the work of fifteen to twenty men with shovels. To provide a harder surface against the effects of rain and thaw, the worst holes were filled with stone where it was available, and the top was gravelled. Broken stone was not necessarily a comfortable surface, but at least it saved many a wagon from being mired in mud or clay. With these techniques a road could be made serviceable except for the heaviest traffic, though its condition depended entirely on the amount of work done upon it. As a precaution against ruts, a law in Nova Scotia required cart wheels to have tires nine inches wide.

The progress of road-building technology varied a great deal from place to place. The Maritimes were approaching the "common" road before Upper Canada was settled. In 1779 a contract for work on the prime Halifax-Windsor route set the following conditions:

. . . a trench to be dug on the upper side of the road, when necessary on both sides; to be rounded in the middle, at the hollow places to be bedded with stones where they are to be had and, when they cannot be had, to be brushed and gravelled; all bad hills, where they may be judged necessary, to have the tops taken off and carried into the hollow to make an easy ascent; the trees and brush to be cut down twenty feet wide on each side of the road from the centre; all short turns and bad hills to be altered where judged necessary.

The contractor was to be paid twenty shillings per rod (5½ yards) for such treatment. It must be noted, however, that this was a road of exceptional importance in the area, and that its condition continued to give rise to frequent complaints.

We have a detailed account of recommended road-building practices in 1841 in a pamphlet written by Thomas Roy, one of the small band of civil engineers practising then in Upper Canada. Laying out roads, he said, should be a systematic procedure, not, as it often was, haphazard or isolated: roadways were public projects, and no local or private interests should be permitted to interfere. If other advantages were equal, the route that was cheapest to build and maintain should be preferred, with due attention to the amount of animal strength expended upon transport; that is, the road should enable the greatest use with the smallest effort, even if its length was increased in the process, for everything still depended on horse and ox power. Experiments in Britain had shown that a load of twenty-one hundredweight could be hauled over a granite stone road at two and a half miles per hour, with an effort depending on the inclination, as follows:

Rate of inclination	Force required
Horizontal	47 lbs.
1 in 229	78 lbs.
1 in 49	115 lbs.
1 in 27	152 lbs.
1 in 22	171 lbs.

Limestone and other less effective surfaces increased the necessary force by about fifty per cent, and there was also variation depending on the strength of the draft animal. To take advantage of this knowledge, Roy remarked pointedly, some understanding of the laws of motion was essential.

Roy felt that the prevailing sixty-six-foot allowance for roads in Upper Canada was inefficient, and necessary only near towns. It required too high a centre, he said, and it increased the cost of maintenance. He recommended an allowance of forty-eight feet—thirty-eight feet for the road, five for a ditch on one side, and the same for a footpath on the other. In the actual construction, he recommended levelling stakes every hundred feet so that proper attention could be given to all-important savings in inclinations by reducing embankments and excavations to a minimum, the most economical method

CROSS-SECTION OF DIRT ROAD

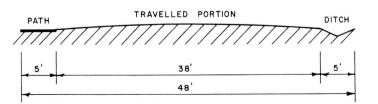

PATH   TRAVELLED PORTION   DITCH

5'   38'   5'

48'

being to make the two balance over a section. Surface cutting was obviously cheaper than deep cutting, but plenty of examples could be pointed out in the province where more money had been spent obtaining a rise of one in fourteen than a good engineer would have spent in getting a result of one in twenty-five. On concession lines as much was sometimes spent on cutting one hill as would have purchased a twenty-mile road allowance in better terrain.

For effective drainage, Roy said, the ditch should be on the side from which flood water flowed towards the road, and of adequate capacity. Frequent lateral outlets should be built, and any spring water along the route diverted; the drainage system might, in fact, have to be carried well beyond the roadbed itself. The centre of the road should be about thirteen inches above both the edge of the ditch and the footpath, and the path should have a shallow water run between it and the road, with a culvert under the road where needed. A smooth surface would prevent the forming of pools of water, and provide as well a better basis for gravel or broken stone. The cost of building such a road in a forest, he said, including cutting the trees and removing the stumps, varied from £220 to £280 per mile, not including the cost of cutting and filling and embankments.

Forty-five per cent of the wear and tear of roads in that day was estimated to be caused by the horses' feet, thirty-five per cent by the wheels of carts and carriages, and twenty per cent by climate and atmosphere. It was apparent, therefore, that the centre should be harder than the edges by a ratio of nine to seven. To give the best results, hand treatment by wheelbarrow and shovel was much more effective than merely dumping from carts. The size and type of gravel should be selected to suit the soil, and one layer of metal was best allowed to settle before another was added.

Roy concluded his pamphlet by recommending the creation of a provincial engineers' department with one engineer in charge of each district. The engineer, and not political or local considerations, would appoint and control superintendents and foremen and instruct them in their duties, for "the degree of perfection to which a road can be brought will ever depend upon the talents, scientific knowledge, and practical experience of the person conducting the work." Roy also made a provocative prediction—that "steam carriages" operating on adequately planned roads could compete with the railway. He believed they might even carry goods at sixteen miles per hour if the road had no curves and no rises greater than one in thirty. Fifty years later, when the first automobile appeared, it was in fact uncertain which would win, the "steamer" or the internal combustion gasoline car. But even by then Roy's "ideal" road conditions were still a long way off.

In 1815 John L. McAdam had introduced in England a method of road construction which is still used with some improvements today. It was very simple, but effective. First a firm base was laid down of large rocks. On top of this were placed layers of progressively smaller stones, culminating with crushed stone at the top; and the whole was bonded together with a mixture of stone-dust and water which hardened into a form of concrete. One of the earliest roads in Canada to be macadamized was the Napanee-to-Kingston section of Ontario's lake-shore highway, a distance of some twenty miles which was transformed in 1837–39 at a cost of $132,000. The construction methods were only an approximation of McAdam's, but for many years that stretch was the exception in and otherwise execrable highway. In the late thirties Yonge Street also was macadamized twelve miles northward from Toronto, and was consequently a good wagon road at some seasons of the year. In subsequent years further pieces of highway were similarly improved, often by turnpike trusts. Rarely, however, was an English standard achieved: the cost of crushing and hauling stone was too

CROSS-SECTION OF
BRITISH MACADAMIZED ROAD

CROSS-SECTION OF
CANADIAN MACADAMIZED ROAD

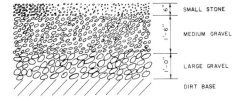

SMALL STONE

MEDIUM GRAVEL

LARGE GRAVEL

DIRT BASE

2 - 3 INCHES OF MEDIUM GRAVEL
DIRT BASE

great for the young economy. As late as 1863 it was said that the only properly macadamized roads in Canada were turnpike trusts outside of Montreal and Quebec; and in the 1890s another critic questioned whether there was yet a mile of real macadam in Ontario outside of a few towns and cities. What was popularly regarded as macadamized road was really nothing more than dirt covered with loose gravel or broken stone, which was quickly thrown by the passing wheels into the centre or the ditch; the wagons then proceeded to wear deep ruts into the exposed surface, and many a wheel was broken. From the March thaws until the end of May such roads were impassable. And for every road so "improved" there were others throughout the nineteenth century which had not yet reached even that stage of refinement.

In one area of road construction Canada did lead the world, however. That was the plank road, which flourished for about two decades. As long as lumber was cheap it was the perfect medium for providing a fast, hard surface which could be easily maintained; and in a land of trees the plank road grew quickly across the countryside. The first of any size was built east of Toronto along the Kingston Road in 1835–36. To James Buckingham, who drove along it three years later, it was a revelation. "I never remember to have travelled so smoothly," he wrote. The planks themselves were hidden by a layer of soil to lessen friction so that at first sight there was nothing strange about the road; but the absence of ruts and pits, and the almost entire lack even of hoof and wheel marks, made this journey very different from any other the well-travelled English writer had taken. So did the low rumbling given out by the carriage passing over the wooden platform. The commissioners in charge of Yonge Street had been equally impressed when the road first opened, and ordered a trial mile built on their own thoroughfare. The expense, they found, was only a quarter that of a

A rare contemporary photograph of a plank road.

stone road, and in view of the unexpected cost of maintaining a macadamized surface they decided to make further use of the new technique.

The Rebellion of 1837 retarded all public works, but after the arrival of Poulett Thomson (later Lord Sydenham) as lieutenant-governor in 1839, the Hon. Hamilton Killaly was appointed President of the Board of Works, and under his direction plank roads were developed in many areas. Within ten years the government had constructed 192 miles of them, and road companies or other private enterprises about 250 miles more. In their heyday plank roads connected Hamilton and Windsor, London and Brantford, Whitby and Port Perry, Cobourg and Rice Lake, and ran in the vicinity of Sarnia, Goderich, and many other Upper Canadian centres of population. The first plank road in Lower Canada was built in 1841 between Longueuil and Chambly. Like many others it was financed by farming out the tolls. The idea spread from Canada to the United States—in reverse of the normal northward flow of innovations in transport—and was widely copied throughout New York State. Sir Richard Bonnycastle, an officer of the Royal Engineers, wrote wonderingly of such an experience in western Ontario: "Fancy rolling along a floor of thick boards through field and forest for a hundred miles. You glide along much the same as a child's go-cart goes over the carpet."

The raw materials of a plank road were readily at hand—all that was needed was a sawmill to cut the timber—but its construction required more care than was usually devoted to the macadamized type.

The width of a good road in this period was sixteen feet, but usually only half the road was planked. The other half, called the turn-off, might be planked later if traffic warranted, but meanwhile it was supposed to be kept in "decent repair" with gravel. If the whole road was planked it was considered better to have two sections of eight feet rather than one of sixteen-foot planks. Sometimes longer planks were laid in groups to provide a wider road so that the wheels of a wagon which had left the road to allow another vehicle to pass would have a projection on which to remount the planking more easily.

Drainage was most important in the construction of plank roads, for dampness from below would cause mildew, and the upper surface was frequently drenched with rain. Engineers therefore called for a ditch on either side of the road at least two feet below the crown, or highest part. The earth base of the roadway was rolled firmly (a portion of the trunk of a large green oak was recommended as a roller), making a hard bed for the stringers to which the planks were to be attached. These stringers were placed under the usual course of the wheels, and sometimes in the centre where the horses would walk, with their upper edges flush with the earth. Their size varied. They could be lighter in sandy soil, for example, than in clay. The road itself was usually sloped about two inches across its width to ensure drainage, and some engineers believed that the stringers on the lower side needed to be heavier

PLAN OF HALF PLANK ROAD

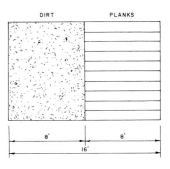

PLAN AND SECTION PLAN OF
PLANK ROAD WITH STRINGERS

PLAN OF FULL PLANK ROAD WITH TURNOFF

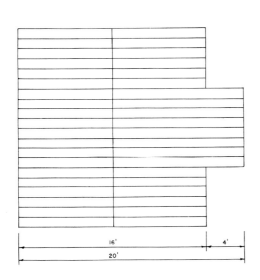

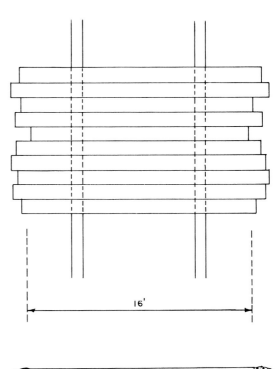

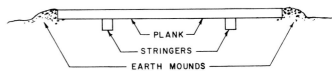

to resist greater pressure—eight inches wide and three inches deep, double the width of the upper stringers. Sometimes stringers were reduced to a six-inch width of one-inch lumber, or omitted entirely, but the effectiveness of the road suffered in proportion. As in masonry, care had to be taken that the ends of the stringers in the various rows did not meet at the same points along the roadway.

Wood was cheap and abundant, and the very best was used for plank roads. White or yellow pine kept sound longest, especially when free from sap, rot, knotholes, and knots. Hemlock, because it had a looser grain, wore away more quickly, leaving hard knots standing out prominently. In the late 1840s beech, maple, elm and other hardwoods came into more common use and in some respects proved better.

The planks, often choice three- or four-inch-thick lumber that is almost unobtainable today, were laid tight against each other and at right angles to the stringers, to which they were spiked securely. Some road engineers embedded them in pulverized charcoal to retard rotting. Normally the planks overlapped the stringers about eighteen inches on either side. Frequently the road edge was deliberately made ragged so that ruts would not form where the boards ended, and usually earth or gravel was crowned up over the ends to give a smoother finish. As a last step in construction the wooden surface was given a light covering of sand or fine gravel. Primarily this was designed to reduce wear from horses' hoofs and wagon wheels; but it also served to lessen the noise and to soak up light rains, decreasing seepage between the planks. Because it was difficult to keep such materials from blowing off, hot pitch was sometimes added.

Grades were a special problem in plank roads, for on hills the wear and tear was at least twice as great. A gradient of one in sixteen presented no particular difficulty but it was considered best to limit grades to one in twenty, the cost of reducing the grade being more than repaid by a decrease in cost of repairs. A certain amount of hill and valley was desirable in any case because it improved drainage, and also because teamsters and stagecoach operators believed that an undulating road was easier on horses than one that was always level.

On a plank road, stages could average eight miles an hour under good conditions, and there were even greater advantages in the hauling of freight. It was claimed that an average team could pull a heavy load thirty to thirty-five miles, day after day, at a rate of three to four miles per hour. Once, on a wager, a team in New York State hauled six tons of iron for twelve miles over a plank road without undue strain, and a common load was one and a half cords of green beech, weighing four and a half tons. The chief danger, it was said, was that the wagon might break from overloading! On the other hand, if the road was poorly built there might be considerable rebound in the planks, which over long distances and at high speeds could seriously harm a horse.

Many of the details of plank road construction are given in a rare pamphlet, *The History, Structure, and Statistics of Plank Roads in the United States and Canada*, published in 1852 by William Kingsford. The author was a great enthusiast for his subject. The spread of plank roads, he argued, benefited more than the traveller. Property in their

neighbourhood took on a new value—up to twenty per cent in the case of good farms—because of the new ease of communication, for the plank road tended to equalize the seasons. It was never closed by mud. Towns that had commonly appeared deserted in spring and fall, when roads were bad or impassable, took on a new life after the plank road had reached them. Their hotels and markets and shops, once busy only during the winter travel season, doubled their year-round business. The neighbouring producer benefited too, because he could get fair prices over the year, and carry three times his former load in half the time: indeed, he could drive to town in wet weather when little could be done on the farm. Even horseshoes lasted twice their regular life!

The economic merits of plank roads as compared with macadamized roads were a matter of considerable controversy at the time. Kingsford said that the former were far cheaper to build and maintain: one man could look after repairs on many miles of plank road in the early years, and even a complete job of relaying would not be overly expensive if the stringers were still sound. Thomas Roy argued with equally impressive sets of figures that in the long run a macadamized road would be the more economical, and time proved him right. The plank road had its place as long as lumber could be obtained for as little as five dollars per thousand board feet in three- and four-inch planks; but gradually the price rose. The roads themselves must have been partly responsible for this upward trend, for they ate up hundreds of thousands of feet of prime wood. Furthermore, at the same time that the costs of road maintenance and extension were increasing, revenue from toll charges was dropping. The railway was making its way across the country, and less long-distance haulage was going by plank road. Caught in an economic trap, the road-owners cut their losses by neglecting repairs. The timbers rotted, leaving dangerous holes, and the ride along them, once described as "smooth as a billiard table," became as unpleasant as any other form of travel. One man reported seeing "a stout elderly lady, when the coach was at a good trot, bumped fairly against the roof by a sudden hole and the shock against the plank at the other side." Where the road ran over sandy soil the timbers lasted longest, and often in very low land rotting planks were replaced for a time as corduroy logs had been earlier, but where gravel was abundant it took over from wood as a material for road construction. Despite the enthusiasm of their champions, plank roads proved to be a curious but very short chapter in the history of Canadian transportation.

Early roads snaked across the countryside. As long as travel was dependent on muscle-power, it was more important to avoid steep grades than to follow a straight line. (Pass of Bolton, Eastern Townships of Quebec, 1837: engraved from a sketch by W. H. Bartlett.)

BRIDGES I

Bridges, beautiful as well as functional, have captured the imagination of the artist and photographer ever since pioneer days when a bridge was apt to be no more than a log felled across a stream, with a crude handrail added. The example on the opposite page was sketched by Mrs. John Graves Simcoe, wife of the first lieutenant-governor of Upper Canada, who used it to cross the Don River when Toronto was still in its infancy.

Bridges have always played an important part in Canadian road-building. About one-thirteenth of Canada's total area consists of inland water; as a result there is scarcely a road of any stature which does not at least once cross a stream, a river, or even a lake. Where the banks were low, the bottom hard, and the current not too swift, the pioneers were able to ford the waterways in their path. Where the distance to be spanned was too great for the technology and economic resources of the times, they depended on ferry boats. Elsewhere they built bridges. And as the country became more settled and prosperous, spindly log scaffolding and corduroy decking gave way to more sophisticated structures of timber and stone.

This is the first of three photographic sections tracing the development of bridges. The others are on pages 140–41 and 214–17.

6

Travelling in Early Canada

SO MUCH HAS BEEN WRITTEN about the difficulties of travel in early Canada that it is hard to realize how extensively the roads were actually used. Not for pleasure, it is true—no one took a trip if it could be avoided, except perhaps on the smooth winter surface—but from sheer necessity. There were no corner groceries or hardware shops in the pioneer settlements. Everything that could not be grown or made at home had to be fetched, often over considerable distance. In the beginning the only way to get about on land was by foot, and even after roads appeared it was still often faster to walk than to battle with a wagon over the bumps and windfalls. Many an early settler carried a sack of wheat on his back thirty miles to the mill, and returned home a day or two later with the flour. To buy a bag of salt, he might have to walk a hundred miles each way. Even a few yards of cloth could involve a prodigious amount of travel and expense. In 1813 Colonel Talbot offered a few of his settlers a bargain: he would supply them with wool if they would have it made into cloth and give him half. Garrett Oakes was one of the men who agreed to do so, and this is his account of what followed:

> I hired a horse and went and got fifty pounds. Here was forty miles travelled. I then hired a horse and took the wool to Port Dover and had it carded, for which I paid $6.25, and returned home, which made one hundred miles more. My wife spun the rolls, and I had made a loom for weaving, but we had no reed for flannel. I then went sixty miles on foot to a reed-maker's, but he had none that was suitable, and would not leave his work on the farm until I agreed to give him the price of two reeds, $6.50, and work a day in his place; this I did, and returned home with the reed. My wife wove the cloth, and I took my half to Dover to the fulling-mill [for final processing]. When finished I had eighteen yards, for which I had paid $34.75 and travelled 140 miles on horseback and 260 miles on foot, making four hundred miles, requiring in all about fifteen days' labour.

With roads as they were, horseback or foot travel was often favoured over vehicle transportation, even for great distances. Records remain of a 2,200-mile horseback journey by missionaries through Upper Canada in 1821–22, a trek on foot through Lower and Upper Canada in 1819, and, in 1822, "a pedestrian tour of 2,300 miles in North America."

Sometimes the cost of inadequate transportation had to be paid in human life. If the nearest doctor was a forty-mile walk away, a man could be excused for hesitating before summoning help—and by the time he did act, it might be too late. Many people who fell ill in pioneer Canada died before a doctor could reach them.

Wagon travel, when it did come to a district, was always an adventure. Accidents were common, and so were breakdowns. No one set out without at least an axe and ropes to get him out of trouble. The wayside was littered here and there with broken wheels and shafts, mementoes of earlier wrecks. Numerous accounts have come down to us from travellers, but one will suffice. Catharine Parr Traill, one of the earliest of Canadian women authors, was bouncing over a corduroy road in a rough cart, the pegged sides of which more than once fell apart. All of a sudden, in the middle of a deep mudhole, the frontboard gave way, throwing the driver into the muck. A little later

Market scenes in Jacques Cartier Square, Montreal, reveal a motley array of farmers' vehicles, ungainly but functional.

a jolt against a pine tree knocked out one of the bottom boards, dumping a sack of flour and a bag of salt pork onto the ground. The driver was scarcely taken aback. The sides needed only new pegs to make them whole; the loose bottom planks were quickly replaced, and off they set again over root, stump, stone, mudhole, and corduroy—a journey that would destroy anything lighter than the rugged Canadian wagon.

Ingenious settlers built their own vehicles. Samuel Strickland, an early settler to the north of Peterborough, described the wagon he built to carry provisions over a newly cut road:

Rough as it was, it was the only vehicle that had any chance of going through without breaking down. The wheels were made of rings, six inches thick, cut off the round trunk of an oak-tree about thirty inches in diameter. Three-inch holes were bored in the centre of these rings of oak for the axle-tree. A strong pole, twelve feet long, was morticed into the centre of the axle for the oxen to draw by, and a small box or rack built on the top of the axle-tree, to which it was fastened by some inch and a quarter oak-pins. The front of the rack was fastened with cord to the pole to hinder it tipping up. Our load consisted of a barrel of salt pork, a barrel of flour, a keg of whiskey, groceries, etc.

Every few minutes the axle would catch against stumps which had

been left standing in the middle of the trail. Then Strickland and a companion would have to lever up the clumsy vehicle with poles until it could clear the obstruction. "And if we were so fortunate as to get along a few hundred yards without being brought up with a jerk by some stump or stone," he concluded, "we were sure to stick in a mud-hole or swamp, instead."

Probably the most impressive of pioneer vehicles were the Conestoga wagons used by the thousands of German settlers who came to south-western Ontario in the first quarter of the nineteenth century. The largest, hauled by six-horse teams, were thirty feet long and carried from six to eight tons. The Conestogas were built almost on the plan of a boat: high at the sides and low in the middle so that on rough roads the load would shift towards the centre, not to one side or end. The body was surmounted by an arch formed of high, broad wooden bows, over which homespun cloth was stretched. The rear wheels were almost as high as a man, the front ones somewhat lower; the timbers were hand-hewn, and the hubs well ironed for protection; and with the running gear painted red and the beds blue, the general effect was one of colour and romance.

The roads of Upper Canada were particularly notorious in the early nineteenth century, but those to the east could not have been much better in the early stages of their development. In the Montreal district many families emigrated to the United States during the 1810s because they had no way to get their farm produce to market; and as late as the 1830s the roads around Montreal were still described as disgraceful. In the Maritimes there were many complaints of routes overgrown or blocked. From Halifax to Truro there were only eleven miles of passable road in 1786; the rest of the way was a bridle path which, according to one traveller, "was generally so soft that even in mid-summer the horses sank to their knees in mud and water, and as each horse put his foot where his predecessor had, the path became a regular

Much Canadian traffic continued to go by water long after the road system was established. W. H. Bartlett, sketching near Montreal in 1837, recorded the convergence of a Durham boat (with sail), a canoe, and a flat-bottomed bateau near a riverfront tavern. In the background, a timber raft sailed downstream, guided by long sweeps.

succession of deep holes, such as one may see in a road made in deep snow."

It says much of the pioneers that they were prepared to leave their own property and travel under such conditions in order to help one another. It was quite common for all the neighbours within a radius of fifteen or twenty miles to gather at one farm to help the owner clear new land of trees and logs—work which might take one man most of the winter, but which could be done co-operatively in a day or two. They would also come to build a newcomer's house or barn, a hundred men often joining forces to raise a large structure. These were social as well as working occasions, and at such bees immense quantities of food and drink were consumed. One wonders sometimes how all the men managed to find their way home along the rough trails and roads, after a day's labour and an evening's festivity. Yet if it was necessary they came back the next day to finish the job.

As the roads improved, more comfortable vehicles made their appearance. In Lower Canada the *calèche* came into frequent use. In New Brunswick the four-wheeled buckboard was a common sight. It had a floor of long, flexible planks which provided some spring to absorb the bumps. In Upper Canada another form of wagon was used, described thus by a local writer, M. G. Sherk:

[It] had wooden axles with a strip of iron above and below, to prevent the wood from wearing away. They were greased with tar, made from the pitch got from the pine trees, and mixed with lard in the winter time, to prevent it from becoming too thick. The tar was kept for the purpose in a special bucket, which was hung underneath the back of the wagon when on a long journey.

The wheels of the old "lumber" wagon were kept in place by linch-pins, which were dropped through a hole in the end of the axle, but as they did not secure the wheel very tightly when the wagon was in motion, they made a rattling noise, which could be heard for quite a distance away. There being no iron wagon springs, the seat was perched on the end of two poles with the ends fastened in the wagon box. This "spring-pole" wagon seat, although high up in the air, was the most comfortable one known.

The development of the elliptical steel spring about 1840 made wagon driving a little more pleasant.

Those who could afford more than an all-purpose wagon bought a buggy. These light carriages sat two people and were often handsomely designed. They had a hood which could be raised as a protection against rain and wind, and their large wheels made them practical

The "jumper" was used for travelling over ground too rough for other vehicles. This one was still in service near Ottawa about 1875.

H. B. Williams' horse-drawn bus ran regularly in the 1850s between Toronto's lakeshore St. Lawrence market and the village of Yorkville, somewhat less than two miles to the north.

over bumps and in mud. In the towns the gentlemen owned fine carriages in which they would take the air with their ladies on a Sunday afternoon. Many of the country vehicles were home-made to the owner's design and, no matter how functional, had a rather peculiar appearance. There were other strange sights on the road, however, among them the travelling menagerie, with its lions and tigers and camels, which provided some of the earliest commercial entertainment in the Canadas.

Increased traffic required some regulation. The first traffic laws were quite simple: they were concerned with the marking of roads in winter, usually by evergreen trees or branches set in the snow along the side. This was the responsibility of the neighbouring landowner, and was so important that he was liable to severe penalties for failure to meet his obligations. He was also expected to clear the road of snowdrifts and fallen trees. Where this was not done winter travellers took the easiest route around the drifts, going over or pulling down any rail fences that stood in the way.

A little later regulations were enacted providing for sleigh bells on harnesses in the winter to warn of oncoming vehicles when visibility was poor. Drivers were ordered to pass on the right and overtake on the left, and to allow half the road to the other sleigh or carriage to prevent collisions. There were penalties for destroying fences and bridge railings. Municipalities could regulate the width of tires and wheels, or forbid heavy traffic on the roads altogether. They could also restrict the driving of cattle, sheep, and pigs over town streets. From the beginning the dangerous driver seems to have been a problem, and long before the arrival of the motorcar there were laws against drunk driving (including men on horseback), "furious" driving (except on the occasional town street set aside to accommodate racing by these antisocial exhibitionists), and the use of improper language when two or more such drivers tangled. There does not appear to have been any law against what was still a major sport in the author's youth in Cobourg, Ontario—"hooking" rides on the horse-drawn sleighs and cutters. To catch them in full tilt was quite a feat; many boys spent their Saturdays at the game.

Another long-standing sport, less respectable but none the less popular among our elders, was cheating the toll keepers. Though it was illegal, the attitude behind it was at least understandable, for the tollgates were a constant source of irritation throughout the nineteenth century. Theoretically the turnpike trusts were established to improve the roads under their care by either planking or macadamizing them, and it was in return for this investment that they were allowed to charge every traveller a fee. In fact, whatever might have been done in the beginning, the toll roads were rarely maintained in good condition— more often the reverse. Residents in the Cobourg area were so angered by the state of the allegedly macadamized turnpike to Port Hope in 1859 that they advertised for one hundred mud scows to be used along its length. Their satirical, venomous announcement continued:

The Company feel that the new mode of conveyance is necessary, as the loss of horses, wagons, and valuable lives in the fathomless abyss of mud during court week was fearfully alarming. Until the completion of the said Mud Scows the Company will continue to exact toll from those who may be so fortunate as to escape alive through the gates.

There were some large turnpike trusts—all the roads in the neighbourhood of Quebec City, for example, were in the hands of one company—but the majority were small. The countryside was a patchwork of local toll roads, in most instances only a few miles long, leading from one village or town to another. Frequently, if not usually, the traveller encountered two tollgates in seven to ten miles of road. William Kingsford gives a description of what one owner considered the most effective arrangement:

The toll-house should extend across the road, so that when the traveller stops to pay toll he should be under the shelter of the roof; and it is desirable that it should be a comfortable dwelling, with cellar and cistern and well and garden, and then the plank road company will be more likely to obtain the services of a civil, respectable, and honest family to tend their gate. The gate should swing—accidents are apt to occur if the gate is made to rise.

Tolls varied, but the traveller at mid-century grew accustomed to rates such as these on the Gore and Vaughan Plank Road outside Toronto:

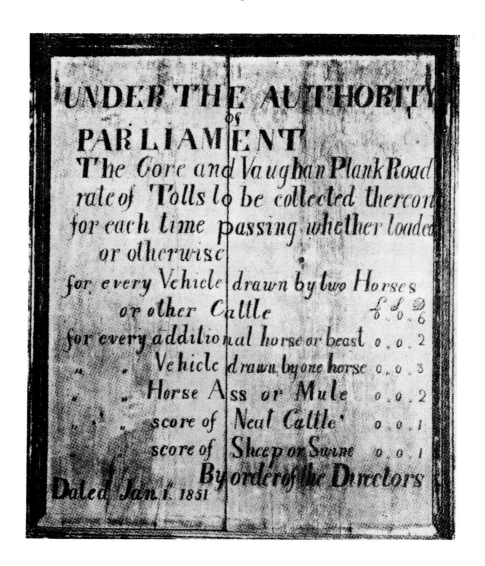

On some toll roads a traveller on foot was charged 1*d*.; on others he went free. A saddle-horse and rider usually paid 2*d*. In later years the charges were higher, and of course in cents. Some curious exceptions were made. People on their way to church were frequently exempt from toll; so were those travelling less than three miles or from one part of their own property to another; and shipments of manure within twenty miles of a city were sometimes not charged. Fines could be levied against toll-road travellers who evaded the gate by going down a side road—a common habit among the more penurious—and these added a little to the revenues, although the difficulty of catching offenders was often considered greater than the proceeds warranted.

"Running" the toll—dashing past the gate at full speed without paying—was common. Even provincial lieutenant-governors were known to have evaded the keeper. Arguments at the gates grew so heated that a law was passed barring profanity on the roads. Other means of protest were more pointed: from Halifax westward, every area has its stories of toll gates being cut down and toll houses burned. Since the roads were not patrolled, most offenders went unapprehended.

According to the terms of their lease many turnpike operators were required to pay the government all receipts above a ten per cent profit on their investment. It was intended that this would form a sinking fund for the eventual purchase of the road, but that end was rarely achieved, for generally profits were too low. Toll roads persisted in

"Cheating the Toll," by Cornelius Krieghoff, captured a popular, if illegal, nineteenth-century sport. Travellers who raced past the tolls without paying would argue that the turnpike owners had defaulted on their part by failing to keep up the roads.

some areas until the first world war, and one at Sarnia, Ontario, continued in use until 1926. The writer has vivid memories of five toll-gates in the Cobourg area in his youth, and of the trouble they caused even towards the end; one of them figured in a manslaughter case after a motorist ran through a tollgate and killed the keeper. Gradually the early toll roads were taken over by municipalities. All that remain are memories occasioned by the survival of such names as "the Old Toll Road" or "the Tollgate Road" here and there. In some areas, however, the idea has been revised by provincial governments as one means of financing expensive projects, such as the $15 million Burlington Skyway overpass in Ontario, which opened in 1958 and within a few years was carrying more than twenty thousand vehicles a day.

Sheer physical effort and determination could carry the farmer to the nearest market town, no matter how bad the roads he had to use. Speedy communication between Canada's scattered cities was a different matter altogether. Today, when we can talk to any part of the world instantaneously by radio or telephone, and fly there in a matter of hours, it is easy to forget what conditions were like little more than a century ago. Overland travel was limited by the strength and endurance of animal muscle. Nothing—neither a person nor a message—could move faster by road than a horse could gallop. And a horse, unlike an automobile, cannot travel all day at high speed. He tires.

The answer was the stagecoach. On the surface the staging system is deceptively simple. A team of coach horses can safely run about fifteen miles (depending on terrain and road conditions) before they become exhausted. At that point a fresh team must be waiting to take over the load for the next leg of the journey. And so on, in relays, until the destination is reached. On the way back the teams return to their original stations.

In practice the system demanded a high degree of organization. A string of first-class animals had to be bought and maintained along the entire route, with all that this entails—stables and stablemen, supplies of hay and oats, provision of water in winter and summer, with adequate allowance for sickness of man and beast. Vehicles and harness had to be kept in tip-top shape, and repairs made quickly. There was never time to waste. Long before the heavy, swaying coach arrived at a stopping point its driver was sounding the alert on his brass horn. At the notes the village woke up, for the stage with its bustle and strange faces and promise of mail was a high point in the routine of life. By the time the steaming horses had been reined to a halt in the cobbled stableyard, a fresh team was standing ready to be hitched, and hot food was waiting for the travellers. Usually there was time for them to step down and take a meal, or at least wash the road dust from their throats, but sometimes the food was simply thrust into the carriage while the horses were changed, the mail dumped out on the floor and sorted, and the old team led away to rest. Then *crack!* went the driver's long whip, and the rattling, bouncing trip began anew.

One of the epics of Canadian travel was a race by stage against time

1841.

SUMMER ARRANGEMENT

BETWEEN

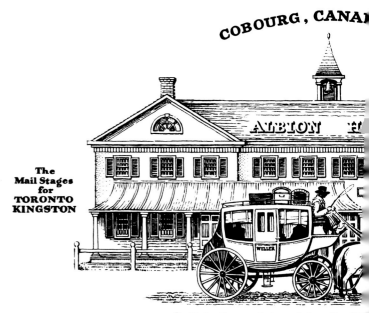

KINGSTON & TORONTO

BY THE

BAY OF QUINTE,

Six times a week each way:

FROM the 1st of May next and during the summer months, the Mail Stage will leave Belleville for Toronto immediately after the arrival of the Bay Steamers, passing through Port Trent, Brighton, Colborne, Grafton, Cobourg, Port Hope, Clarke, Darlington, Whitby and Pickering.

GOOD FOUR HORSE COACHES

(Entirely new,) with steady experienced drivers, going through from Belleville to Toronto in twenty four hours, and from Cobourg to Toronto by day light.

REDUCED FARES.

Belleville to Toronto,	- -	120 miles.	- -	£1 0 0
Cobourg to Toronto,	- -	72 do.	- -	0 10 0
Port Hope to Toronto,	- -	65 do.	- -	0 10 0

The above line of Stages will leave the General Stage Office, Toronto, for Belleville, every Sunday at 10 o'clock, A. M. and every Monday, Tuesday, Wednesday, Thursday, and Friday at 5 o'clock, P. M. after the arrival of the Steam Boats from Niagara and Hamilton.

Strangers will find a great advantage in taking this route; by leaving Kingston (the Capital) in a Steam Boat, they have a fine view of the country forming the Bay of Quinte, fast rising into importance since the late alteration of the Seat of Government, and taking the Stage at Belleville, will pass through the above named townships, which for fertility of soil and density of population will yield to none in the Province, thus reaching the city of Toronto at 8 o'clock P. M.

WM. WELLER,

Proprietor,

Cobourg, April 28, 1841. tf36

N. B.—A Steam Boat leaves Kingston going up, and Belleville going down the Bay, every morning, (Sundays excepted.)

Stagecoaches followed the principal routes still used between cities today. Many a hotel owed its prosperity to their schedule, as indicated by this facsimile of an 1840 business card, redrawn by Frank C. Taylor.

from Toronto to Montreal in the winter of 1840. The Governor-General, the future Lord Sydenham, made the trip himself, it was said to reprieve a prisoner sentenced to hang. The driver was William Weller, owner of Upper Canada's most prosperous stage line. Before they left, Weller bet $1,000 that he could cover the distance in thirty-six hours. It was February and the roads were fast. The Governor was tucked into a warm bed on a sleigh, the start was timed, and, whirling a twenty-foot lash with a sound like a pistol-shot, Weller began the trip. Every fifteen miles his hostlers stood ready with four fresh horses and hot food. Without a stop for sleep the stagecoach owner drove through the night and next day—and just thirty-five hours and forty minutes after leaving Toronto pulled up his steaming horses outside the Montreal office. The prisoner was saved and Weller received, in addition to his fee and side-bet, a handsomely engraved watch from the appreciative Governor-General.

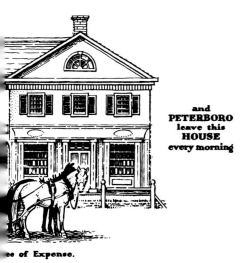

and
PETERBORO
leave this
HOUSE
every morning

se of Expense.

The distance covered was 360 miles. Weller had averaged ten miles an hour, but his actual travelling speed had been closer to fifteen miles an hour. That doesn't sound like much today, but the regular schedule allowed four and a half days for the trip, with time for meals and overnight stops to sleep. Thirty to seventy-five miles was considered a normal day's run. Stagecoach travel, even at its best, was usually so uncomfortable that the passengers needed resting-spots almost as much as did the horses.

It was many years before the calibre of coaching reached the level described above. The popular *calèches* of Lower Canada were the first vehicles to be used as public conveyances. They had just enough room for the driver and two passengers, who by the end of the trip would be spattered with mud or choked with dust, depending on the weather. This was called travelling by post. *Calèches* were used as early as 1780 on the route from Quebec to Montreal, which was divided into twenty-four stages; a similar vehicle was available for hire on the Niagara portage as early as 1798. Elsewhere the "stage-wagon" was seldom more than an open cart, with a plank floor and rough board seats. Gradually these were displaced on the main routes by enclosed carriages; but in more outlying regions they continued in service. Often they were hired specially for a trip when no regular service was provided. In winter, sleighs were used—frequently enclosed and sometimes equipped with small stoves.

In the Maritimes regular stage service began at the turn of the nineteenth century over the well-travelled road between Halifax and Windsor. In 1821 a weekly trip was advertised which took one day in winter, and by mid-century the stage routes had spread in every direction from Halifax. William Weller's counterpart in Nova Scotia was C. H. Belcher. Six handsome greys, we are told, pulled his coaches of "royal blue or chocolate brown, with undergear and wheels a lemon chrome and black striping." Belcher offered daily service in both directions between Halifax and Kentville, and three trips a week between Halifax and Annapolis.

In Lower Canada the main stage route ran between Quebec and Montreal. The trip took two long days, from 4 A.M. till 8 P.M., even with four-horse teams, but by 1813 a coach was setting out on it six days every week. Other routes in the early nineteenth century led from Montreal to Albany, and from Quebec to Boston. There was a weekly run from Montreal to the Long Sault on the Ottawa River, at which point passengers took to boats for the rest of the way to Hull.

The development of stage service in Upper Canada was slowed by the state of the roads. In 1817, when the Kingston Road was opened, Samuel Purdy began a line from Toronto to Kingston, where it met another coming from Montreal; but the route was so difficult that his schedule was maintained only during the winter. In the spring Purdy closed shop and summer travellers had to use lake boats. Ten years later the trip could be made in rough stage-wagons, but even then there were no bridges across some of the larger rivers, and no guarantee that ferries would be operating. It was not until Weller's brightly-coloured coaches—light yellow, drawn by six bay horses—began operating in 1830 that Upper and Lower Canada were linked by a regular fast service. By that time another company was offering scheduled two-day trips to the London area, and there was a well-established line from Toronto to Niagara. The Niagara Portage, one of the most heavily travelled stretches of road in the country, had been turned into a pleasant one-day stage jaunt, with a three-hour stop to view the Falls.

Most stagecoach operators carried mail as well as passengers. This was in fact one of their most important functions: Weller said as much by naming his company the Royal Mail Line. To the operators it meant a guaranteed income from the government, and it could make the difference between profit and loss in a year. To the people along the route it meant the beginning of what we now consider normal postal service. Until the stages began a few hardy couriers maintained a skeleton delivery service across the provinces, but weeks or months might pass between their visits to the points which happened to be on their routes. By far the majority of letters were entrusted to private travellers to be left at stores and taverns on their way, where they might be picked up by the persons to whom they were addressed, or by another passer-by to be carried the next step on their journey. The system worked by good luck and good will rather than by organization, and it was not at all speedy. A merchant in Montreal, writing to an associate in Toronto, could scarcely expect an answer in less than three weeks. The stagecoaches cut that time in half. With them, for the first time, Canadians had a fast, regular, dependable way of communicating with one another. They knew that a letter mailed in Quebec on Monday would reach Montreal two days later, and Toronto in a week. If they had to travel, there was no longer need to beg a ride with a friend or hire a carriage especially for the trip: they could take the stage, secure in the knowledge that it would leave at the time scheduled, and arrive at its destination on the appointed day. Accidents happened—mudholes, or a broken wheel, or a washed-out bridge could delay the coaches— but operators like Weller and Belcher staked their reputation on reliability and did their level best to maintain it.

They had little else to sell; certainly not comfort. The stage-wagons used on minor routes in particular could be miserable conveyances indeed. One traveller has left a graphic description of his experiences on such a trip in November, 1846, from Hamilton to Niagara, a distance of fifty to sixty miles. The stage was a lumber-wagon with a canvas covering. They set out at 6 P.M. on a dark, rainy night—eight people, including the driver, crowded onto its rough board seats, among them a young child who cried most of the night. The roads were so bad that on several occasions they had to get out and walk through the clinging mud. Finally, when all were exhausted, the stage entered Niagara as the sun rose. Such conditions were by no means unusual. It was said that in the early days of the Kingston Road some settlers walked the 160 miles from Toronto to Kingston, and arrived a day ahead of the stage!

The grander coaches which eventually ran on all main routes were still heavy, clumsy contrivances, hung on leather springs which absorbed little of the shock of travel over the rutted, pitted dirt roads of the day.

A judge, who presumably had to make use of them frequently on the business of the courts, described one:

> The body was closed at the front and back and covered with a stout roof. The sides were open, but protected by curtains that could be let down if rain came on; there was a door at each side fitted with a sliding window that could be lowered or raised as the weather was fine or stormy. There were three seats inside, each of which was intended for three passengers; those on the front seat sat with their backs to the horses, those on the back and middle seats faced them: the back seat was the most comfortable. Outside there was the driver's seat, and another immediately behind it on the roof; each of these would hold three persons. The best seats in fine weather were those on the outside of the coach, as they commanded a good view of the country on all sides. . . . At the back of the coach body was the baggage-rack for trunks, which were tightly strapped on and protected by a large leather apron. Lighter articles of baggage were put on the roof, which was surrounded by a light iron railing. . . . The whole affair was gaudily painted.

Not everyone agreed that the back seat was the most comfortable. Some preferred the centre seat, which was hung on leather straps. Others preferred to sit beside the driver and pump him for stories about the district through which they were passing. In some less elaborate models the passengers had to climb in through the window; these coaches were designed without doors to keep out the water when they had to cross through rivers and streams which were not bridged.

Some of the hazards of the road of those days are still with us. We can all sympathize with a young lieutenant who planned to spend the time between Annapolis and Bridgetown reading, "but was incessantly interrupted by a prosing little woman, eighty years of age" who had taken a fancy to him and persisted in small talk of every kind. Other dangers, fortunately, have disappeared. The roads were so rough that Catharine Traill said she usually emerged at the end of a journey "black and blue," and another writer, after a particularly hectic trip, remarked that "broken heads" were not a rarity on such occasions though fortunately none of his companions on that journey had been hurt. The chief cushioning was the press of passengers: the coach lines worked on the principle that there was always room for one more, and that a traveller squeezed tightly between two other people had less chance of being hurled from his seat when the vehicle pitched into a hole.

At times the driver would call to the passengers, "Gentlemen, a little to the right," or "a little to the left," at which request all would throw their weight to the appropriate side to keep the vehicle upright, and sort themselves out afterwards when the balance had been restored. On other occasions everyone might disembark and walk beside the coach in order to lighten the load while the driver navigated a particularly difficult stretch. They might even have to tear down a fence and help pry the heavy vehicle out of the mud, or give a hand dragging it up a slippery slope. Occasionally there were complete breakdowns. A typical ride was recounted by Lieutenant-Colonel B. W. A. Sleigh, one of the many British soldiers who left a record of their experiences in North America. In April, 1852, he was travelling in one of the new, large Concord-type coaches, with seventeen other passengers and much baggage, when the leather springs gave way over the heavy Nova

Scotian roads; the break was repaired but occurred twice again, with the result that he had to walk five miles in pouring rain.

In good weather and on a good road, stage travel undoubtedly could be pleasurable. Even Mrs. Traill, despite her bruises, championed the coaches as the best that could be devised. The worst time to venture on the roads was in the spring, when they were heaved by frost and washed out by freshets. Even in summer, however, the stage lines often ran on a reduced schedule, and many passengers made long stretches of the journey by boat, taking to the coaches only at rapids and waterfalls. The best time to travel still was winter. Then wheels were replaced with runners, and the travellers, if even more crowded because of their bulky clothing, at least had a relatively smooth passage. Anna Jameson saw such a mail-sleigh at a tavern in Oakville, Ontario:

It was a heavy wooden edifice about the size and form of an old-fashioned lord mayor's coach, placed on runners and raised about a foot from the ground; the whole was painted a bright red, and long icicles hung from the roof. This monstrous machine disgorged from its portal eight man-creatures, all enveloped in bearskins and shaggy dreadnoughts, and pea-jackets, and fur caps down upon their noses, looking like a procession of bears on their hind-legs, tumbling out of a showman's caravan. They proved, however, when undisguised, to be gentlemen, most of them going up to Toronto to attend their duties in the House of Assembly.

Now and then the stage was held up and robbed, but highwaymen were never as serious a threat as they were in other times and places. Passengers were in much greater danger from their own coachmen. The drivers were tough men who enjoyed mastering the spirited team of horses drawing the lumbering coach. Most of them also enjoyed a drink, and took one when the opportunity arose while the horses were being changed. After a number of stops the rides were apt to be rougher and the possibility of upsets greater. In the winter of 1848 a drunken stagecoach driver drove his coach over the edge of the bank into the St. Lawrence River. Other accidents were caused by recklessness on the part of the driver, or from fatigue. One coach turned over in a ravine after its driver fell asleep and the unguided horses wandered off the road in search of water.

The coach stops were always convenient to an inn for the benefit of the passengers as well as the driver. It was alleged from time to time that some drivers were in league with certain innkeepers to delay the journey so that everyone would have time to spend a little more money at the inn. Whether that is true or not, the roadside inn was as important a part of coaching as the vehicles themselves. It offered ample meals, if not always perfectly prepared, with plenty of meat, pies, puddings, and fruit. When the passengers stopped for the night there was also rude accommodation—frequently, it must be admitted, shared with bed-bugs or worse. Many pioneer inns were neat and clean, but others were far from well-kept.

The greatest enemy of the traveller was the road itself. No matter how perfect the organization of the stage line, there was no guarantee that its coaches and passengers would arrive at their destinations without misadventure. The coming railways promised greater speed, safety,

Stagecoaches came in many forms, from open wagons to heavy, clumsy carriages "looking for all the world as if elephants alone could move [them] along," as one traveller wrote. In winter, they donned runners. This one ran between Toronto and Kingston in 1829. Some cold-weather models had wood stoves to keep the passengers warm.

comfort, and punctuality. As they spread across the country in the eighteen-fifties and sixties, the long-distance stage lines gradually disappeared.

At a grand dinner in honour of the opening of the Cobourg and Peterborough Railroad in 1854, William Weller, the stage proprietor, made this contribution:

I know why you have called upon me for a speech—it is to hurt my feelings; for you know I get my living by running stages, and you are taking the BIT out of my mouth at the same time as you take it out of my horses' mouths. You are comparing in your minds the present times with the past when you had to carry a RAIL, instead of riding one, in order to help my coaches out of the mud. But after all I am rejoiced to see old things passing away and conditions becoming WELLER.

7

Brigade Trails and Gold Rush Roads

IN THE SAME YEAR that William Weller was good-naturedly accepting the eventual demise of his coach lines, the first road was built in what would become British Columbia. It ran from Victoria to Esquimalt and was constructed by a company of sailors, for Esquimalt—till then the centre of a farming district—had been pressed into service during the Crimean War as a supply base against Russia for Britain's Pacific Squadron. In 1854 the southern tip of Vancouver Island was the only settled portion of the province. Its few hundred residents had managed, in Fort Victoria and the surrounding countryside, to follow a way of life remarkably close to that which they had left behind in England, but they were only just beginning to require trunk roads. The previous year the Council had appointed a committee to lay out a route from Victoria to Sooke, a settlement west of Victoria on the coast, and Governor James Douglas had initiated a liquor tax, over strong opposition, to pay for road-building and schools. The Council subsequently appropriated £500 for roads and bridges, and drew the money from the Hudson's Bay Company, which by the terms of the Royal Grant of 1849 was responsible for all costs involved in colonization.

Across the Strait of Georgia the mountains of the mainland thrust upward, the most formidable challenge to transportation Canadians would meet. Here even the rivers—the natural highways of eastern Canada—were too swift, too often interrupted by falls and rapids as they raced through the narrow gorges, to serve as traffic arteries. "We had to pass where no human beings should venture," Simon Fraser wrote after descending the river that bears his name. In the roaring canyon north of Yale he had picked his way along on old Indian trail, clinging along the perpendicular sides to a flimsy ladder of lashed poles which swayed dizzily outward over the chasm as he moved. Fraser and other early explorers, among them Alexander Mackenzie and David Thompson, were searching for a fur-trade route to the Pacific. This obviously was not it.

Three years later, in 1811, Thompson did find a route by way of the Athabasca and Columbia Rivers, but the cost and effort required to travel it were prohibitive. In the end, the mountains forced the fur-traders to abandon their canoes and establish separate communication lines on the West Coast. Most trading posts were built at the mouths of rivers where they could be reached easily by ocean-going boat. The few in the interior were located close to rivers but nevertheless were approached by land. Over the passes and along the narrow valleys the fur traders carried their supplies on the backs of packhorses.

The earliest of these routes was in the north. Fort McLeod, the first European outpost in British Columbia, was founded in 1805 on McLeod Lake, where it could be supplied from the prairies. The next year the North West Company established Fort St. James, which was to be its administrative headquarters, on Stuart Lake. To reach the new post by river was possible but roundabout and difficult. It was far easier to carry supplies eighty miles overland, even in the beginning when it had to be done on foot or by dogsled.

In the south a far more extensive communications network was developed. Once a year trade goods and food for the interior were shipped up the Columbia River to Fort Okanagan. From there the trail stretched 225 miles, up the grassy Okanagan Valley and on to Kamloops. The annual brigade might easily consist of two or three hundred packhorses, many of them unbroken. Smaller trains, and on at least one occasion a herd of cattle, continued through the rocky wilderness as far as Fort Alexandria, halfway up the Fraser to McLeod. On their way south the brigades carried the year's return of furs to be shipped from the Columbia to London.

The Okanagan route to Kamloops was well suited for pack trains, but the boundary settlement of 1846 placed Fort Okanagan and the lower Columbia River within United States territory. About the same time, Indian wars erupted in the area. The Hudson's Bay Company was forced to take a hard second look at the Fraser as a potential means of entry to the interior. Difficult as it was, it was at least British. The first stretch, to the point where the river turned north, was navigable. From there, eastward traffic faced the steep slopes of the Cascade Range. Two routes were discovered through this barrier. The first was so difficult that it was abandoned in its second year. The other was used for more than two decades. From Fort Hope, near the head of navigation on the Fraser, it crawled up the valley of the Coquihalla River and over Mansons Mountain to the headwaters of the Similkameen, then turned northward to Kamloops; from there brigades could follow the old trail further to Alexandria. On this track, near the summit of the pass through the Cascades, Simon Fraser's son Paul was killed when a tree thundered down upon him as he slept. At another point a traveller in 1859 reported seeing the skeletons of sixty or seventy packhorses. "My horse," he reported, "frequently shied at the whitened bones of some one of the poor animals, who had broken down in the sharp struggle with fatigue and hunger."

The brigade trails, dangerous and difficult as they were, sufficed for the fur companies, and until 1857 no one else was much interested in the interior of British Columbia. Then gold was discovered on the Fraser and the Thompson. In the spring of 1858 the rush began.

Thousands of men flooded into the Fraser Valley. The first came from California, where the gold finds of '49 were petering out. Many of those who followed had never done a day's work outside an office: they had thrown down their pens on an impulse to seek quick wealth. Veterans and tenderfeet, of diverse colours and nationalities, they pressed up the valley from its mouth, some in boats, others on foot or on horseback or crowded in wagons—whatever they could find. The snowfall that winter had been unusually heavy, and now the river was flooded; scores were drowned in its swift current but the rest kept going until the magic word—"Gold!"—was shouted near Hope, where the Fraser turns north. By September ten thousand men were rocking and washing the banks between there and Yale. Some found as much as $50 a day. But this was placer gold, tiny fragments of the precious metal washed down by the river from some unknown source. Where was the mother lode? The next spring the rush moved north in search of it. By Christmas it had passed Lillooet and Soda Creek and Alexandria, and had reached the Quesnel River. A thousand men were pan-

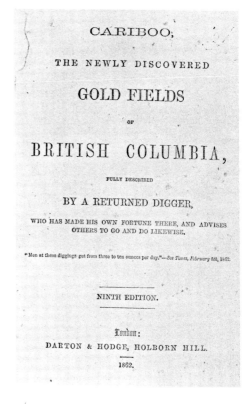

CARIBOO,

THE NEWLY DISCOVERED

GOLD FIELDS

OF

BRITISH COLUMBIA,

FULLY DESCRIBED

BY A RETURNED DIGGER,

WHO HAS MADE HIS OWN FORTUNE THERE, AND ADVISES OTHERS TO GO AND DO LIKEWISE.

"Men at these diggings get from three to ten ounces per day."—*See Times, February 8th, 1862.*

NINTH EDITION.

London:

DARTON & HODGE, HOLBORN HILL.

1862.

ning around Quesnel Lake and in the golden gravel of Horsefly River. And then, early in 1861, they poured over into the Cariboo. That was the Eldorado of the day. In a single year it produced $2.5 million in gold. From all over the globe thousands of men, and some women, converged on this once-lonely spot.

How did they reach the Cariboo? The first stage was relatively easy. Steamboats crossed regularly from Victoria—still the commercial centre of the colony—to Fort Langley, near the mouth of the Fraser; in July of 1861 a steamboat reached Fort Yale. The new head of navigation had only been a fur post; overnight it became a brawling town. From Yale the pioneers used canoes where possible. Where they couldn't they followed the treacherous Indian path. It had not changed since Fraser described it half a century earlier:

We had to pass many difficult rocks, defiles, precipices, through which was a kind of beaten path traversed by the natives, and made possible by means of scaffolds, bridges and ladders, . . .

For instance, we had to ascend precipices by means of ladders composed of two long poles placed upright with stocks tied crossways with twigs; upon the end of those others were placed, and so on to any height: add to this that the ladders were often so slack that the smallest breeze put them in motion, swinging them against the rocks, while the steps leading from scaffold to scaffold were often so narrow and irregular that they could scarcely be traced by the feet without the greatest care and circumspection; but the most perilous part was when another rock projected over the one we were clearing. The descents were, if possible still more difficult: in those places we were under the necessity of trusting our things to the Indians—even our guns were handed from one to another. Yet they thought nothing of it; they went up and down these wild places with the same agility as sailors do on board a ship.

The miners had no native guides to help them. Quite the reverse—132 were killed by Indians during 1858 before peace was finally enforced. They had to carry their supplies over the ladders on their backs. In the most dangerous places they took off their hobnailed boots and clung with their toes to the rock walls. Some abandoned their canoes and followed the upper banks on foot. Simply because there was no alternative, large numbers of them finally took time off from the gold hunt to cut a mule trail from Yale to Spuzzum and Boston Bar. As far as Spuzzum it followed the first leg of the old Hudson's Bay brigade trail—the one to Kamloops which had been abandoned because it was too dangerous.

As the rush moved upstream in the summer of 1858, plans were laid for a new route which would avoid the difficulties of the Fraser as far as Lillooet. To build it Governor James·Douglas enlisted the support of five hundred miners who had left the field broke and discouraged: if they would supply the labour, he would provide food and transportation. The new way followed a chain of lakes which lay in a narrow, heavily wooded valley between high ranges of mountains. On the long, deep lakes boats could be used; but in the stretches in between, one of them twenty-nine miles long, trails had to be cut. Work began at the head of Harrison Lake at a point the volunteers named Port Douglas, after their sponsor. In the first four days they opened ten miles of wagon road. Later, as they moved farther away from the base at Douglas, the pace slowed. Mules were scarce that summer—

any that could be obtained were at the diggings—and supplies had to be carried to the head of the road slowly and expensively on the backs of the builders themselves. Nevertheless, by fall a trail had been cut and bridged. The following year the road was improved with the help of the Royal Engineers. In 1861 further work was done by soldiers and private contractors, who finally produced a wagon road which a naval officer who used it said "would be no discredit to many parts of England." On one short stretch there was even a primitive railway. The track consisted of wooden rails covered on two sides with angled iron, and the rolling stock consisted of a tiny flat car which ran by gravity down the grade one and one-half miles from Anderson Lake to Seton Lake, and was hauled back up by horses. By 1862 wagons and stages were operating along the entire land route and connecting with steamships on the lakes.

But the Douglas-Lillooet Road was far from satisfactory and never popular. The trip across it was long and tiring, with its several changes between land transportation and the steamers, which never seemed to be at the docks when they were wanted. It was, moreover, a slow and costly route for freight. In one court case arising over a delay in shipping goods, witnesses testified that from twenty-five to sixty days were required to transport merchandise between Douglas and Lillooet. The judge said he considered forty days a reasonable time. A sarcastic and greatly overdrawn description of the trip was published by critics who sought an alternative:

For scores of miles, the Cariboo Road was blasted and carved out of the mountainside. The men who built it came from many parts of the world. Some were on their way to the gold fields. Others had tried their luck and failed, and now were earning the money to return home— or to try again.

Travellers are assured that they can get through from Douglas to Lillooet in from 20 to 40 days, and at a cost of $150, so if you are determined and will be humbugged, the fault is your own. To prevent your straying from the "high-toned and elegant route" read the following directions:

Take a splendid steamer at New Westminster for Harrison River. There hire elegant Indian canoes to pole you over the rapids, or walk along the pebbly shore, wade four sloughs and swim one small river to reach a high-toned propellor which runs at the speed of two miles per hour (wind permitting). No close confined cabins on board, but pure wholesome air on deck, with the privilege of sticking your nose in the cook's galley to warm it without extra charge. Twenty-five hours will take you to the mouth of the Douglas slough, where she connects with capacious canoes, fare $2.00 to the edge of the ice near Douglas rancheries (smallpox there, but don't hurt white man, only kills Indians) then foot of snow. The little lake being frozen over, walk around it to Lillooet Lake —scenery delightful. Then catch another elegant and high-toned steamer if you can: if you can't, wait a day or two—meals only $1.00. When the steamer "Toots her horn" get aboard and rest yourself on the open deck for four hours: weather moist or air keen. Reach Pemberton. Good meals there for $1.00 each, beds 50 cents, crawlers gratis (smallpox blankets carefully washed). Rest a day there and foot it again for 24 miles to Anderson Lake; catch a steamer again if you can; rest again on the open deck going over the lake; foot it again for 1½ miles or take a ride on the railroad car (?) to Lake Seaton; catch another splendid steamer, if she is in repair, for Port Seaton; foot it again 3½ miles to Lillooet; rest there three or four days (the smallpox is played out, Indians are killed); then swim your horse across the Fraser (if the ice permits) to Parsonsville; then run up Pavilion mountain to help circulation. . . .

Where the mountain dropped away, the road ran on timbers. Its 385 miles were completed in three years.

From Lillooet, as the final lines of this passage suggest, it was possible to continue up the Fraser along a trail notorious for grades which rose steadily four thousand feet to the summit of Pavilion Mountain, and

dropped the same amount more rapidly on the other side. No matter how exaggerated the criticisms might have been, clearly this was not the route by which the riches of Cariboo could be best exploited.

An alternative route already existed in embryo. Despite all the effort poured into the Douglas Road, most traffic to the upper Fraser still followed the river. There was, of course, no wagon road, but trains of sure-footed mules could make the trip to Cariboo in about a month. The Royal Engineers had blasted a new and at last acceptable trail from the head of navigation at Yale up to Spuzzum. From there the pack trains proceeded up the Fraser and Thompson canyons to Spences Bridge, where they headed back to the Fraser by way of Clinton, and thence to Quesnel. Each animal could carry 250 to 400 pounds, lashed to its back with a diamond hitch. From sixteen to forty-eight mules might be used in a single train, with about one man (including the cook) to every eight animals. When this trail was opened the freight rate from Yale to Quesnel dropped from one dollar to forty cents per pound. But the final stretch from Quesnel to the mines was so bad that two English visitors reported seeing scores of dead pack animals, "some standing as they had died, still stuck fast in the deep, tenacious mud."

As long as the Cariboo gold held out, the miners formed the greater part of British Columbia's population and provided the major part of its wealth. But neither resource could be properly exploited, or even administered, without an effective means of communication between the gold fields and the seaboard five hundred miles to the southwest. To achieve this end Governor Douglas proposed nothing less than a wagon road, eighteen feet wide and adequate for stagecoaches, from Yale to the diggings and dance halls at Barkerville.

Seventeen miles above Yale, the road clung to the walls of the Fraser Canyon.

The 385-mile Cariboo Road was one of the wonders of its age. In places it ran close to the water's edge; elsewhere it was so high that the mighty Fraser looked like a tiny stream. Here its path had to be blasted out of the living rock; there it hung to the precipice on a balcony-like set of trestles. Where rock-falls had broken the canyon wall it spanned the gaps on log trestles or masonry fill. Near Spuzzum, seeking an easier grade, it crossed the canyon itself on the first suspension bridge built in British Columbia. Yet the whole route was completed in three years at a cost of about two million dollars.

The road ran from the log houses and frame saloons of Yale up the fearsome Fraser canyon. North of Boston Bar it passed over Jackass Mountain, named after the many mules that had dropped to their deaths while negotiating the old trail, in places only inches wide, along the side. At Lytton it left the main stream, turning up the Thompson and Bonaparte Valleys on a detour which, after some very difficult stretches, offered a relatively easy passage through waist-high grass to Clinton. From there up steep hills to a stretch of timber where the road had to be corduroyed for many miles. At Carpenter's Mountain came infamous beds of miry clay, then a swift descent back to the bank of the Fraser at Soda Creek. Here the river was navigable, and the next leg, to Quesnel, could be made on the *Enterprise*, a steamboat which had been built on the spot by Gustavus Blin Wright, one of the contractors, from locally cut wood and parts that were packed in pieces on muleback over the trail from Yale. From Quesnel the highway rose again, over a sixty-three-mile route along Lightning Creek and through the Devil's Canyon to Barkerville.

Construction began in May 1862. A few of the more difficult stretches were built by the Royal Engineers, but elsewhere the work was done by private contractors under almost insuperable conditions. Funds were so short that in lieu of immediate cash some builders were given the right to collect tolls—once the road was completed. (Where they got their working capital in the meantime was their own business. Blin Wright, who had contracted to build the section south of Alexandria in

CARIBOO ROAD

Great Bluff on the Thompson River, 88 miles from Yale—still 297 miles to the gold fields.

return for a five-year right to collect tolls over it, tried to borrow money from the innkeeper at Williams Lake. When he was turned down he changed the road to a more difficult course, simply to avoid the lake and to pass instead a roadhouse at 150 Mile which had a more accommodating owner.) Pack animals were almost impossible to find in many areas, and boats were short elsewhere. So was labour: many men deserted to the mines as soon as they got their wages, and in the end much of the work was done by Chinese who laboured long and hard and troubled their employers only by their desire for hard-to-obtain pork. Mosquitoes and horseflies attacked in swarms, rain was frequent, accommodation was primitive, and government grants hard to obtain. Beyond all these "normal" difficulties of early road-building was the terrain. Walter Moberly, one of the principal contractors (and before that one of the most intrepid explorers of routes through the mountains) wrote this description of the route as it was in 1863, before it was completed, in a letter to Sandford Fleming, the government engineer-in-chief.

Commencing at Yale the head of Steamboat Navigation on the Fraser R. to Sailors bar, a distance of 7 miles, built by the Royal Engineers—very heavy blasting, wall building and side cutting—Grades easy—cost $32,000. From Sailor's Bar to the Suspension Bridge built by Thos. Spence. Eight Miles in length Contract price $36,000—heavy blasting, side cutting and much timbering —Wire Suspension Bridge nearly 200 ft. span, private Speculation—built by J. W. Trutch at a cost of $35,000. Government Grant charter for seven years— Toll ⅓ of a cent per lb. on all goods & a small toll on animals and carriages— Charter redeemable. From Suspension Bridge to Boston Bar Eleven Miles built by J. W. Trutch—four miles of which is Tolerably Easy other seven miles the very heaviest description of work—high wall building, an immense quantity of blasting, heavy side cutting and timbering—cost about *$150,000.* From Boston Bar to Lytton thirty-two miles built by T. Spence at a cost of *$104,000.* Country very much cut up steep ravines, a great deal of heavy side cutting, much timbering and blasting. From Yale to Lytton along the Fraser the road passes through a rough, rocky, broken mountainous and sterile valley. On the whole of this section there are only three or four gardens.

From Lytton to Cook's Ferry on the Thompson, a distance of twenty-four miles the road runs along the south bank of the Thompson river through a rough, rocky, broken and barren Valley—was built by W. Moberly at a cost of *$130,000* —much heavy blasting, side cutting and timbering. The Thompson is here crossed by a ferry (a wooden bridge 500 ft. in length will most probably be built this next season at a cost of 15 to *$18,000*) and for a distance of Eight Miles to Venable's Creek the road was built partly by W. Moberly and partly by the Royal Engineers at a cost of *$32,000.* Much blasting and side cutting—road nearly level. From Venable's Creek to Moberly's Camp a distance of 4 miles was built by Thomas Spence, *12 feet* wide (*Govt. funds short*) cost *$9000.* From Moberly's Camp to Clinton (The Junction) a distance of forty-one miles— built by W. Moberly—Generally through an Easy Country—it passes partly along the north bank of the Thompson and the west bank of the Bonaparte rivers, through Maiden Creek Valley (Glen Hart) to Clinton—not much blasting some timbering & side·cutting—Cost *$70,000.* Country sterile and barren— little or no timber One or two small poor farms—a good deal of bunch grass but much Alkali.

For a short time there was brisk competition between the Yale and Douglas roads as routes to the north. Freight-haulers, stage lines, roadhouses, and other interests advertised flamboyantly in the newspapers

One of F. J. Barnard's stagecoaches leaves the terminus at Yale on the four-day journey to Barkerville and the diggings.

Ox teams hauled loads of food and other supplies to the free-spending miners.

of Victoria and New Westminster, but it soon became evident that the new route had all the advantages. It was more direct, and a traveller could proceed along it without the delays and trans-shipments which marred the land-and-water passage from Douglas to Lillooet. Soon the Cariboo Road carried the bulk of traffic.

Along it miners tramped in red flannel shirts, corduroy trousers, and top boots. If they had been lucky they carried gold in their knapsacks on their way south, but more often than not they returned to civilization as penniless as the day they set out. Trains of pack mules, up to fifty or sixty in single file behind the white bell-mare, still appeared on their way to the hills. But now there were heavy freight wagons as well, two or three perhaps linked together behind a dozen yoke of oxen, signalling their approach with jingling harness bells as they carried fresh eggs and milk and vegetables, boots and shovels, champagne and pianos, to the stores and saloons of the roaring towns which had sprung up near the mines. Prices in Barkerville and Richfield and Cameronton—where as much as $50 had been paid for a pair of boots and $10 for a single egg—dropped suddenly.

At one point five steam-powered tractors were brought over from

Climbing Jackass Mountain on the road to Cariboo, 1868.

Scotland to compete with these bull-trains; but the first time one was used it proved no faster and considerably more expensive than old-fashioned ox-power, and the scheme was speedily abandoned.

For those who could afford it there was a regular stage service. Great Concord coaches, with red and white bodies, their running gear and wheels a brilliant yellow streaked with black, rumbled along at eight miles an hour behind teams of four or six horses. F. J. Barnard's "express" coaches made the trip from Yale to Barkerville in four days, travelling night and day with fresh teams waiting every twelve to fifteen miles. In the deep black-leather boot in front they carried an iron safe for transporting treasure from the mines. On the seats inside, rocking to and fro on the leather springs, sat officers and prospectors, saloonkeepers and gamblers, hurdy-gurdy girls of the dance halls and, often, the wives or sweethearts of the gold-seekers. Many a bride made the journey to Eldorado, hoping to return home with a golden fortune; but most of them, some with the men of their hearts, fill a grave in the small hillside cemeteries of the Cariboo. They and their times have been remembered at Barkerville in a modern reconstruction of the old boom town, built as a gold rush memorial and museum.

Even on the coast, British Columbia road-builders faced gargantuan obstacles and overcame them with surprising speed. A British traveller made this sketch in June, 1888, as work began to clear the Douglas firs which blocked the extension of a Vancouver street. In August he returned. "Not a tree or stump stood there; but a splendid plank-road was laid down, a wide side-walk being on either side, and rows of comfortable dwellings, side by side, for half a mile beyond."

If the nineteenth-century traveller was fortunate he might even catch a glimpse of one of the fabled camels of British Columbia. Twenty-two of these animals had been imported in 1862 by way of San Francisco from Texas, where the United States Army had used them as carriers. The Cariboo Road was still then a trail, and Frank Laumeister, one of the packers, believed they would make his fortune, for camels can carry two to three times as much weight as a mule. On the relatively easy Douglas-Lillooet road he was proved right, but when he transferred his camel trains to Yale disaster struck. The soft-footed animals, used to sand, were crippled by the rocky terrain, and the scent of camel enraged and terrified the mules used by other packers. After several accidents the camels were withdrawn. Some were disposed of in Victoria, but others were loosed in the Thompson Valley and roamed the interior ranges until about 1905. It was said that an unwary traveller might find himself eating wild camel, masquerading as beef, in one of the coaching inns along the route.

There was only one Cariboo Road. Some other grandiose plans were made, and failed. In 1861, for example, after gold was discovered at Rock Creek, road-building was started along the year-old Dewdney Trail from Hope to the Similkameen. By the time 25 miles were opened, Rock Creek had been abandoned as the miners streamed north to richer fields. Two attempts also were made to reach the Cariboo directly from the Pacific, from Bella Coola and Bute Inlet. A mule trail was actually opened from the former place, but never improved; the second attempt ended after a construction crew was massacred by Chilcotin Indians.

With Confederation came a change in the emphasis on road building in British Columbia. Through the 1860s the main concern was in gaining access to the gold fields; in the 1870s the government concentrated on opening trunk routes through the settled areas of Vancouver Island and the rich farming district of the lower Fraser, and to a lesser extent in the few populated portions of the interior. In 1874 work began on a road from Ladner, on the Pacific coast, up to Hope: river traffic, even along the broad lower Fraser, apparently was growing inadequate to meet the needs of expanding settlement. This road was notable in that, in glaring contrast to most others in the province, the path it followed was too low. The first nine miles east from Ladner crossed delta land which at high tide was covered with salt water. The builders began by digging trenches on either side of the roadway and piling the dirt in the middle to form a dike. The top was then levelled and corduroyed. The ditches were used to drain the land and for local travel by canoe.

During the 1870s British Columbia devoted nearly forty-five per cent of its provincial revenues to the construction and maintenance of roads and bridges, and in this and the following decade the road network gradually spread both on the mainland and on Vancouver Island. Yet much of the work was still impeded by inefficiency and lack of overall planning. Funds were allocated piecemeal as pioneers settled beyond reach of existing roads and demanded new ones. Much of the building and maintenance was left to statute labour under elected foremen; and many roads were only a few miles long, designed solely for local use. In parts of British Columbia the road appropriations were considered something of a dole—and indeed for many families who provided the unskilled labour, the appropriations were the major means of livelihood. In Canada's bustling westernmost province, in short, road conditions through this period were much the same as in the east.

8

Prairie
Trails

WHILE EASTERN CANADA AND BRITISH COLUMBIA were opening trunk routes and colonization roads, the prairies remained for the most part fur-trading country, the domain of the Hudson's Bay Company. The only settled area lay in the Red River Valley, where Lord Selkirk founded the Assiniboia colony in 1812. After initial discouragement that community had taken firm root, and by the 1830s it was an important source of food for the trading-posts scattered across the west. It was also their principal link with the rest of North America. Between the prairies and Upper Canada lay hundreds of miles of rock, trees, muskeg, and lakes in the Canadian Shield. There was, however, a relatively easy passage south from the settlement to St. Paul, the terminus of steamboat travel up the Mississippi River. And so the first well-used highway in Canada's west ran into the United States.

Along this route the normal vehicle was the Red River cart, named for the district in which it had originated. This light wagon—really little more than a box on wheels—was made entirely without metal. Its axle, shafts, body, and two big wheels were all of hardwood from the prairie riverbanks. Instead of the iron hoops which were normal in the east, the wheel rims were bound with tough green buffalo hide to protect them from wear. Travellers usually carried a supply of this *shag-a-nappi* (rawhide) on long trips for emergency repairs. The wooden construction had the advantage that at rivers and streams—none of which were bridged for many years—the wheels could be disconnected, strapped to the bottom of the body, and the whole affair floated readily across. Perhaps the most striking characteristic of the Red River cart was its noise, a creaking and shrieking of elm hubs against maple axles which could be heard for great distances across the flat grassland.

The carts were made in quantity at White Horse Plains, a few miles west of Fort Garry (Winnipeg). They formed the first manufacturing industry in the west. Light but strong, easily repaired by anyone with an axe, saw or auger (the city of Moose Jaw is said to have got its name because a Red River cart was mended there with the jawbone of a moose), they were the perfect vehicle for crossing the prairies. On the route from St. Paul they were often hitched into "brigades" of four to six carts, each wagon with its own horse but all in a line managed by a single driver. After 1845 these brigades travelled in great trains, as many as five hundred Red River carts in a gigantic file moving slowly up the valley to Fort Garry. Normally the day's journey began with the sun, halted at mid-day when the heat became intense, and resumed in the cooler evening until dusk. Fifteen to twenty miles were considered a good day's progress; a trip could take three weeks to a month. The charge for freight was around sixteen shillings per hundred pounds, and each wagon could carry five hundred to a thousand pounds.

The existence of the Red River colony and the trail to St. Paul revolutionized communications in the Canadian west. The long canoe route from Montreal via the Lakehead had been abandoned when the Hudson's Bay and North West Companies merged in 1821; fur-trade transportation thereafter was based on Hudson Bay. It still depended

The Red River cart was the birchbark canoe of the prairies—light, easily made, easily repaired, perfect for travel across the grasslands. This photograph was taken in 1873 at a Métis traders' camp at the Elbow of the South Saskatchewan.

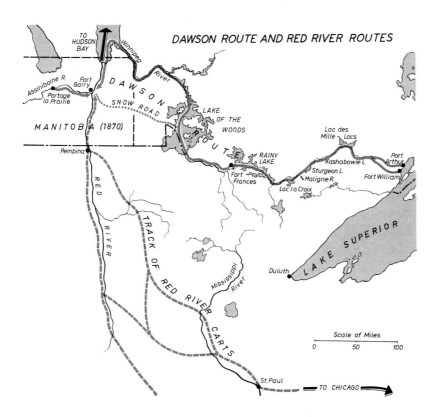

DAWSON ROUTE AND RED RIVER ROUTES

on water: goods from England were shipped to York Factory and taken by boat down the Nelson or Hayes River to Norway House at the head of Lake Winnipeg. From there the distribution network spread in all directions: south by Lake Winnipeg and the Red River to Fort Garry; east by the old canoe route to the Great Lakes; west and north by the Saskatchewan, Athabasca, and Mackenzie Rivers. It was a complex and costly system but it worked on the basis of annual shipments from England, largely because it was controlled by a single company.

The growth of the Red River settlement provided a supplementary supply of food, and one which could be drawn upon in winter when Hudson Bay was locked with ice. About 1829 the flow from York Factory south was reversed: a winter road was begun northward from Norway House, and while it was not at first successful, in the mid-thirties this route was developed into a regular post-road with storehouses and stables. The growing trade between St. Paul and Fort Garry added to the importance of this northward traffic. Distribution from Hudson Bay never ceased, but shrank in proportion. When Sir George Simpson, governor of the Hudson's Bay Company's northern district, visited the Red River colony, even he preferred to go by train to Chicago and from there overland, rather than by his company's traditional northern route.

In 1853 the United States extended postal service to Minnesota, and from then on most of the mail to the Red River colony also travelled by way of St. Paul. This was a great improvement over the Hudson Bay route, which was sometimes responsible for a letter's taking eight months to travel from England to Fort Garry. But with both mail and supplies flowing northward from the United States, real concern began to develop whether Assiniboia, completely isolated from other British settlements in North America, could be kept under the Union Jack.

An old Hudson's Bay Company photograph shows the back-breaking labour involved in hauling a York boat over as simple an obstacle as a beaver dam.

A Geological Survey of Canada crew on the Red River in 1858. The canoe still played a vital role.

Annexation of the area by the United States seemed almost inevitable. To ward off this feared outcome an all-Canadian postal service to the Red River was begun in 1858. Twice a month in summer the mail was carried from Toronto to Collingwood on Georgian Bay, from there by steamer to Fort William, and then by canoe to Lake Winnipeg. In winter monthly dogsleighs travelled along the wild northerly shores of Lakes Huron and Superior. Michael Labatte, one of the couriers, colourfully described his experience on this arduous route:

> Made trip from Penetang' to Sault and back (300 miles) with sleigh and two dogs in fifteen days—snow three feet deep. Once made trip in fourteen days. Dig hole in snow with snow-shoes, spread spruce boughs, eat piece cold pork, smoke pipe and go to sleep. Once five days without any food but moss off rocks.

This service was abandoned after two years, but Canadian fears of losing Assiniboia were to have more lasting results. Two missions were sent west in 1857 to investigate the suitability of the prairies for colonization. One, sponsored by the Imperial government in London, was led by Captain John Palliser. He and his men spent three years in careful surveys of an area where the buffalo still roamed in hundreds of thousands. At one point Palliser was forced to walk for several hundred miles after his horse was killed while he was travelling alone. His report was cautious: he believed a large area of the southern plain was too dry for farming—a prediction which was ignored for decades until the so-called Palliser triangle turned into a dust bowl during the drought of the 1930s. On the vital subject of communications Palliser was even more pessimistic. He felt that an all-Canadian route to the west was impossible, and that settlers would have to continue to go round the Shield through the United States to St. Paul. At any time it is undesirable for one country to have to depend upon another's goodwill for transportation, but given the then-current Canadian suspicions of United States expansionism this was particularly unwelcome advice.

The other expedition, consisting of Simon J. Dawson and Professor H. Y. Hind, was sent by the Canadian government. Perhaps reflecting the mood of its sponsors, it was far more hopeful. Hind recognized an arid belt, but a far smaller one than Palliser. Moreover Dawson, an engineer, believed it was possible to open an up-to-date route by land and water to the new area from Fort William, following the old Kaministiquia Portage. He foresaw the many carrying-places turned into roads with corduroy across the swamps and bridges across the rivers. The larger lakes would be crossed by steamboats. Where the water was too shallow for boats he planned to raise it with dams.

Dawson's optimism won the day, and the route was named after him. The first stretch of dirt road was opened west of Lake of the Woods in 1868, and another segment was completed forty-eight miles west from Port Arthur, via Kakabeka Falls ("the Niagara of the North") on the Kaministiquia River. During the winter of 1869–70, bridges were built over the two largest rivers in that portion. Then, unexpectedly, the barely-started Dawson Road was needed to meet a national emergency. Along the Red River, the first Riel Rebellion had broken out. Fourteen hundred troops under Colonel Garnet Wolseley were ordered to proceed west to put down the insurrection, but because they were in uniform they could not pass through the United States. They had no choice but to turn road-builders and join Dawson's crew in opening the trail before they could operate as soldiers.

Wolseley's army set out from Port Arthur on July 1, 1870. Before it lay a wilderness inhabited by wandering tribes of possibly hostile Indians. Rivers, lakes, rocks, and scrub timber were broken only by burned-out clearings left by forest fires. Guides were scarce, blackflies overwhelming. Over that route they had to transport two field guns, ammunition, equipment, tools, camping and cooking gear, and provisions for sixty days. The soldiers quickly learned to carry heavy loads over portages with tump lines. But this was no ordinary crossing with canoes. They were travelling in York boats, made of planks, with shallow keels, six oars, and sails. Each boat weighed seven to nine hundred pounds: to drag them across a portage was no easy matter. A road ten feet wide had to be cut and roughly corduroyed. Fortunately the area abounded with poplar, which has a smooth bark, and when it rained the slippery logs were almost as effective as greased rollers. Even so it took eight men to haul each of the lighter boats, and thirty or

Wolseley's troops begin the long portage around Kakabeka Falls, dragging their heavy York boats on rollers up the slope. William Armstrong, an English engineer, accompanied them to the Red River in 1870 and recorded the scene.

forty were frequently necessary to move the larger ones over steep places. At one of the most difficult points a rocky cliff had to be surmounted: a ladder of felled trees was built and the boats hauled up by ropes. Had one of the ropes snapped, several men would certainly have been injured and the craft itself smashed.

All in all there were forty-seven portages, some more than a mile long. Over all of them the men usually had to make ten trips in order to carry the great quantity of supplies, so that a mile of carriage necessitated nineteen miles of walking before all the stores and equipment were transferred. Officers and men laboured from 5 A.M. to 8 P.M., with only short breaks for breakfast and dinner of salt pork, biscuit, and tea. Yet no one became sick, and by August 23 the first soldiers reached Lower Fort Garry.

The Dominion Government now took over the road and underwrote its completion as Dawson had planned. In July 1872, less than two years after Wolseley's expedition, Sandford Fleming travelled from Port Arthur to Fort Garry in nine days. His party used wagons on the first stretch to Shebandowan Lake, where Indian guides were waiting with canoes to take them through to Lake of the Woods; there they returned to wagon transportation. Even the Indians had a relatively easy time, for steam tugs towed the boats over the open water. At its peak the Dawson Route boasted corduroy roads with horses and wagons along its major portages, and steamboats and tugs on twelve or fourteen sizeable lakes like Kashabowie, Mille Lacs, Sturgeon, la Croix, Rainy, and Lake of the Woods. Dams on the Maligne River raised the

water as much as twelve feet, and there was a lock at Fort Frances. Way stations were built of logs for overnight stops. During 1875 some two thousand persons travelled this route to the Red River. Many were immigrants planning to build new homes on the prairies. It was not a very comfortable trip—140 miles of rough roads and 310 miles of boat travel in all kinds of weather, beset by flies and mosquitoes, with crude sleeping accommodation and the danger of food running short. But it was a bargain. The fares from Fort William to Fort Garry were advertised at $15 per adult and less for children; Fleming estimated that his passage cost the government at least ten times as much. In the end, the Dawson Route contributed to its own death. The builders of the Canadian Pacific Railway made good use of it in carrying men and supplies, but when they were finished the road was abandoned almost overnight.

To complete the story we have an account by Grace Lee Nute in the Hudson's Bay Company publication *The Beaver* of an eight-day canoe trip in 1953 in search of relics of Dawson's dream.

At the first portage [after Port Arthur] we found the remains of a dock, a corduroy road of enormous pine logs, and a cleared space. Dawson established overnight hostelries on strategic portages, and I wondered whether this was one of them. If so, it had a romantic setting for overnight guests, with its pines, a thundering cataract, and gray cliffs. At the next portage we found the head of a steam tug's boiler and a small propeller. . . . At the very next portage we found another dam and the tug. . . . Dawson's old tug lay in shallow water, resting on its side in the remains of an engine mount and the scattered timbers of the sunken craft. . . . Here was one of the actual vessels that had plied up and down this stretch of the river carrying passengers and freight at least seventy-five years ago.

West of Winnipeg the prairie stretched to the horizon—a natural road-

Kashabowe Station on the Dawson Road, painted by William Armstrong on a return journey in 1911. Steamboats shuttled between portage points.

Contemporary cartoons suggest that the soldiers managed to keep a sense of humour during their three-month journey through the wilderness beyond the Lakehead.

way uninterrupted by dense forest or jagged rocks. It was good country for horses. These animals were not native to the Americas but had been brought to Mexico by the Spaniards: some of them had escaped, turned wild, and spread northwards up the centre of the continent, ahead of the white man. To the Indians they were an invaluable ally in hunting the bison on which plains life depended. A brave had a much better chance of making a kill from the back of a horse than he had on foot. When it was time to move camp to follow the roaming herds, the horse could also be made to pull the family belongings. For carrying goods the Indians used a *travois*, a triangular frame of two long poles which were lashed to the animal's sides and dragged behind it on the ground.

Later bison-hunters, knowing of the wheel, used a Red River cart instead. They were Métis, of mixed French and Indian descent, who lived in the Red River Valley around Fort Garry and St. Boniface. They or their fathers had been fur-traders, and the farming life was not for them. In great bands they would set out across the prairies, under quasi-military discipline, to find and kill their shaggy quarry. As years passed and the eastern prairies became hunted out, the Métis cast out further and further from home. As many as 1,200 Red River carts might be used on a single hunt to carry supplies on the way out and to haul the loads of hides and meat on the way home. The hides would be sent to St. Paul to be made into coats and robes by the Americans; the meat would be dried, pounded into paste, and mixed with fat and berries to form pemmican, the long-lasting, nutritious staple food of the fur-traders and explorers.

No roads were made on the hunt: the pursuers cut across open country or followed the tramplings of the herd. Gradually, however, wagon trails began to appear, criss-crossing the prairies. Although the Hudson's Bay Company distribution system was based on rivers, it was often more direct and economical to cut across land. Definite routes became established between the scattered fur-trading posts. With time the passing cart wheels ground deep ruts into the dry plains soil. When these ruts became too deep for easy passage the Métis carters would break a new trail beside the old one. On some of the most heavily travelled routes as many as twenty ruts ran side by side to the horizon.

There were thousands of miles of cart trails. Few were straight. They ran from one ridge or headland to another, shifting direction as their makers searched for the combination of wood and pure water that was necessary for survival. One of the most important trails ran from Fort Garry nine hundred miles northwest to Edmonton House by way of Fort Ellice on the Assiniboine and Fort Carlton on the North Saskatchewan. Brigades of Red River carts could carry pelts, provisions, and trade goods between these two major centres of the fur trade in six weeks if they were travelling fast. Another route ran south from Fort Carlton to the site of Saskatoon and from there to Fort Benton, a United States outpost on the Missouri River which gave access to the entire Mississippi system. Edmonton was also linked by trail to

A Métis brigade of Red River carts sets out to hunt buffalo. (Painting by William Armstrong.)

Fort Benton. One of the major trails in southern Saskatchewan ran from Fort Qu'Appelle (which in turn was linked by trail to Fort Garry) southwestward to Fort Walsh, and then south to the United States. Where it crossed the big bend in the Moose Jaw River—a favoured location because of its plentiful wood and water—this trail met others leading to Wood Mountain, Cypress Hills, Red Deer Forks, and by way of the Elbow to Fort Carlton. Many of these fur-trade cart trails are still followed by today's main highways—proof of the good judgment of their first users.

Along the trails carved by the carters there came the settler, heading west. Three decades before the prairies themselves were opened for settlement they were being crossed by adventurers and families on their way to British Columbia. The Saskatchewan Trail from Red River to the Rockies was used by many famous men. Among them were the artist Paul Kane and Sir George Simpson, the latter on his way to inspect Hudson's Bay posts on the Pacific. Simpson's party once made the nine hundred miles from Fort Garry to Edmonton in twenty-two days on horseback, with the baggage following in carts. On the way they overtook a large party of immigrants under James Simpson, a famous leader of such groups. The twenty-five families were headed for the Columbia River Valley. Each family had two or three carts, and horses, cattle, and dogs. Several births occurred en route.

The "Overlanders" of '62 made the journey in search of quick wealth

The settler retained the Red River cart as an all-purpose vehicle though not, it appears, without occasional difficulties.

in the gold fields on the other side of the Rockies. This large party, which included women and children, undertook the crossing of the prairies at a time when the voyage around Cape Horn—or via a railway in plague-ridden Panama—was long and extremely arduous. Some of them were victims of fraud, lured by the advertisements in British and Canadian newspapers of a company that promised first-class stage-coach accommodation from St. Paul to British Columbia. Except for a head office in London where fares were collected, the company had no existence.

The Overlanders did not learn of the hoax until they reached St. Paul. There, however, they were able to take a stage to Georgetown on the Red River, where a steamer carried them to Fort Garry. The boat was late in starting and a faulty steering gear made the trip up the curving river slow and arduous—but worse was to come.

At Fort Garry the expedition bought horses, oxen, Red River carts, pemmican, and flour. The carts, equipped with semicircular canvas covers, cost £8 to £10 each with harness. The pemmican, so hard that it had to be chopped with an axe, cost sixteen cents a pound. On June 2 the group set out—ninety-six carts in all, each with an eight-hundred pound load—led by a guide who knew the country. The men were well armed, for Indians to the west were threatening to tax anyone crossing their land.

The first step of the journey was a tiring eleven-hour ride to Long Lake where they found the water almost undrinkable. This was to be

a recurring danger in a land of shallow alkali lakes. Here the company stopped to organize itself. A captain was appointed, and a committee formed to make and enforce rules. Camps were to be triangular in shape, with the carts in rows along the sides, the animals tethered inside, and the tents on the outside; six guards, two on each side, were to keep watch all night. Each day the camp was roused at 2.30 A.M. and was under way by three. A stop for breakfast and they set off again. Dinner was at 2 P.M. Four hours later they stopped again to make camp. At two and a half miles an hour a good day's journey amounted to thirty miles. In common with most western expeditions, there was no travel on Sunday.

"It was an inspiring sight," wrote one who saw it, "to view the train from a distance, winding its way round picturesque lakes or slowly extending out on the lovely landscape gorgeous with wild flowers of every hue." Those closer at hand were perhaps less inspired. The route led them from the Hudson's Bay post at Portage la Prairie due west to Fort Ellice, a two-week trip. Here lay the Assiniboine River, and the only ferry was a scow drawn by rawhide ropes at both ends; since it could take no more than a single cart and ox at one time, and that with difficulty, the crossing was slow and tiring. Now the route turned northwest. There was a steep hill which caused many accidents, and another primitive ferry at the Qu'Appelle River, this time with the added aggravation of a fifty-cent toll levied by the Hudson's Bay Company on every oxcart.

The next night, at Gulch Creek, the guide deserted. The party pressed on. The trail was well marked to Fort Carlton, but the days were hot and mosquitoes swarmed. The water supply was meagre, salty and sulphurous. At Fort Pitt the weather broke. For eleven days it rained; everyone's clothes were wet and tattered but there was no rest. Bridges had to be built across flooded rivers—eight of them in all, constructed with trees from the valleys. Trunks forty to a hundred feet long were cut and hauled across the torrents by men wading in mud and shoulder-high water; then corduroy was laid down across these supports and the carts moved one step further toward their goal. At Fort Edmonton came a brief period of relaxation. The Overlanders put on three minstrel shows in return for kindnesses. Then they sold their carts, traded their oxen for packhorses, and set out towards the swift rivers, canyons, storms, and precipices of the Rockies—but that is another story. Those who eventually reached the Cariboo had spent four months en route.

In the 1870s true settlement of the prairies began. At the beginning of that decade the Hudson's Bay Company transferred its vast holdings in land to the Canadian government. The area of Canada was multiplied six-fold by the transaction. Almost immediately survey crews began staking out townships and mile-square sections from Fort Garry westward, and homesteaders were promised free land. The first flow of immigrants doubled the prairie population before 1880. Thousands of them bumped their way along the rough trails in "prairie schooners": four-wheeled wagons with a high, curving cloth cover under which the women and children rode with the family possessions, protected

Along a dirt trail cut by the buffalo hunter and trader through the prairie grass, a train of covered wagons winds its way westward. Four-wheeled oxcarts belonged to the International Boundary Commission mapping the Saskatchewan border.

from sun and storms. These vehicles were related to the Conestoga wagons of Ontario but were more simply made and considerably lighter: a team of four or even two horses could pull them in comfort. Often oxen were used instead. For hauling heavy loads the pioneers used bull trains—heavy, huge freight wagons, coupled in a line and pulled by six to twelve pairs of oxen hitched tandem. With good leaders and firm ground twenty tons could be hauled ten miles per day. The driver, called a "bullwhacker" or "skinner," walked or rode alongside his train, plying his long, cracking whip and equally explosive vocabulary. The best area for such heavy cartage was in southern Alberta, and many of the loads made their way to Fort Benton for shipment through the Mississippi Valley.

Easy as prairie roads were to open compared with those in the east and far west, they were not without difficulties of their own. "I had always understood the prairie was so beautifully smooth to drive over," wrote one disillusioned emigrant, Mrs. Cecil Hall; she found it in reality pocked with gopher holes and hills and cracks, more like an uncultivated field back home in England. "If your carriage is heavily weighted it runs pretty easy, but woe betide you if driving by yourself—you jump up and down like a pea on a shovel," she added.

There were still no proper roads across the prairies in the 1880s. Particularly dreaded were the patches of adhesive mud known locally as gumbo. In autumn the trails were sometimes dry and smooth, but conditions in general were summed up lucidly by W. F. Rae, author of *Newfoundland to Manitoba*. The traveller heading west from Fort Garry in 1881, he wrote, would find no road. Only the general direction was indicated.

It may be said, indeed, that each traveller makes his own road. If he be aware of the direction which he ought to follow, he chooses the part of the prairie where the ground is best fitted for driving. Nothing is easier than to drive over the stoneless and springing turf of the virgin prairie, and if the traffic be not too great an excellent "trail" is made by the passage of successive vehicles. But when the traffic is heavy and continuous and holes are formed in which water settled and the soft mould resembles a mass of tenacious mud, then following the "trail" is a weariness to the flesh of man and beast.

Trains of freight wagons were often detained for days just west of Winnipeg because of the difficulties, Rae said. He advised any prospective British emigrant to "practise crossing a newly-ploughed field for hours together with a horse and cart and pitching a tent at the end of his journey. Let him arrange so that there are frequent ponds in the field, these ponds being at least five hundred yards in width, having an average depth of four feet and a muddy bottom."

Scattered along these rough trails Rae found "hotels"—in reality rude log houses, sixteen by eighteen feet or thereabouts, which the traveller shared with the innkeeper and his family. Food generally was limited to fried salt pork, bread, potatoes and tea; eggs and milk were rarely available. Some innkeepers charged tolls of twenty-five cents simply to cross a bridge near their dwellings, and the flow of traffic was so great that they could gouge $50 a week from passing immigrants in this way before the government stepped in to stop the practice.

After the opening of the Canadian Pacific Railway in 1885 immigrants usually travelled by train as far as they could before setting out overland. In the 1900s they came by the thousands. Mary Hiemstra, who is still living, has told in *Gully Farm* of her childhood experience as part of the Barr Colony, one of many emigrations from England organized by land-owning companies. Knowledge of Canada began with a week-long trip from the Maritimes in a colonist coach, dirty, overcrowded, and often without food. The rails ended at Saskatoon, where tents were still more common than houses. Here the party bought wagons. Their destination was Battleford, 120 miles up the North Saskatchewan. Most of the colonists were city folk who had never before seen farm animals or implements or even been in a wagon.

Two tracks wound along the path of least resistance, skirting sloughs full of brown brackish water, hills and depressions, and clumps of bushes and scrub that in the west went by the name of trees. The wagons creaked and swayed over badger holes and gopher mounds, and some of the passengers walked in preference to the everlasting bumping. At night they camped wherever they were. The slough water was dirty, but the wigglers at least were strained out by a cloth. Bread ran out, but one to the manner born showed them how to make bannocks of fried dough. Some colonists were met on their way back, thoroughly disgusted. Even in late April and early May, snow and slush were often on the ground, and at times the trail was lost for days

Prairie homesteaders on their way to a new life.

at a time. Battleford proved to be merely another collection of tents pitched on rough, rocky land. But, like many another western place, it was the gold at the end of the rainbow for thousands.

In winter the settlers could find a smoother ride over the snow by sleigh. There were no trees, however, to break the cold winds. Drifts formed, and in a blizzard it was easy to lose the poorly marked trail and wander onto the open prairies. Neverthless many a pioneer made his way across the country without too great hardship in a closed sleigh which was little more than a packing case on runners with a stove inside and a slit at the front through which he could see and manage the reins. For some the sleigh continued to be a temporary home, even in the bitter cold of winter.

Over the years the cart trails that had been turned into colonization roads became country roads of a rough and ready nature, improvements consisting mainly of filling the sloughs with logs and dirt and sometimes digging drainage ditches. Most farmers on the prairies lived in isolation on their quarter-sections. The road carried them to the nearest town, with its general store and post office and livery stable. The road also led to the railway, and there they hauled their wheat in open wagons to be stowed in elevators before being shipped east.

In the chief towns the bases of several modern main streets were laid. Portage Avenue in Winnipeg and Jasper Avenue in Edmonton, and many other prairie city streets, are extraordinarily wide for one reason: when their boundaries were being set the rule of the road was to move to the side if the ruts in the middle were too deep, and that took space. Pontoon bridges were built over the Red and Assiniboine Rivers, but they were frequently damaged by carelessly operated barges; since the cost of repairs was great, navigators were ordered to blow their whistles loud and long so that the bridges could be dismantled in time.

The gradual evolution of the western road is clearly shown along the Edmonton-Calgary Trail. Much of it began as Indian paths. Then in 1873 Rev. John McDougall cut a trail from Fort Edmonton southward to his mission at Morley, which today is headquarters for the Stoney Indians of the Morley Reserve. Two years later, when the North West Mounted Police founded Fort Calgary, the trail was extended eastward to their post. By the time the Canadian Pacific Railway reached Calgary in 1883 it was a busy wagon-road. In that year the first regular mail service was inaugurated between the forts, and the first stage established. It left Edmonton each Monday at 9 A.M., stopped at Peace Hills (Wetaskiwin), Battle River (Ponoka), Red Deer Crossing, and Willow Creek (in the Olds area), and arrived at Calgary on Friday. Soon four types of stages were available, varying from freighter wagons with or without seats to open and closed coaches of a more comfortable but in no sense luxurious type. The fording of streams was always a problem, but in the late eighties modern vehicles were sometimes brought in from London, Ontario, at a cost of $300 each, and these at least kept the passengers dry. By that time freight hauling totalled some 2,500 tons a year. The first official survey of the Calgary-Edmonton Trail was made in 1886. The surveyor had modern ideas on roads. "In view of the great traffic and immense travel which some day may be done this way," he said, "my intention was to make the road as straight as the actual direction of the trail between the two extreme points would allow." His foresight has paid dividends. Today Alberta Highway No. 2 still follows closely the old Calgary-Edmonton Trail. Modern drivers cover the distance in a little over three hours.

At the turn of the century Manitoba—or at least the southeastern corner of it—had been a province for thirty years and had begun a road programme of sorts under its own Department of Public Works. The remainder of the prairies was still five years away from provincial status: it was known as the North West Territories, and roads, as such, were administered from Regina. Dr. G. D. Stanley, in an article in the *Alberta Historical Review*, told of pioneer medical service to the cattle ranchers and scattered homesteaders of southern Alberta. The trails he followed in his horse-drawn democrat (a light wagon with two seats) wound in and out across the prairies, over hills, around or through sloughs, through rivers and streams. An emergency call might involve a forty- or fifty-mile trip each way. Where the trails stopped, the doctor went by landmarks by day and by stars at night. The bridge over the Highwood River was the only one between Calgary and Fort McLeod; elsewhere the traveller had to ford the streams. If the water was high this meant a swim for the horses while the doctor stood on the seat— or swam himself. In winter he crossed on the ice, which occasionally gave way. As farming developed and villages increased and graded roads multiplied, some of the difficulties lessened, but there were still pot-holes of sticky gumbo to be crossed, and other problems:

Another obstacle, which the pioneer doctor had to face daily and often, were the barbwire gates which gave the trails admission to the fenced-off ranching or farming districts, through which the trail wound itself to find the shortest and safest route over which it could pass. Further, if the doctor's team happened to be still in the broncho stage of breaking, and the doctor preferred to have them for use during the balance of the journey rather than walking, each gate demanded the tying of those bronchos to the gate post twice, first on the high side while the gate was being opened, and again on the off-side while it was being closed; and, believe me, the stretching of the barbwires to close the gates was some job at times.

So wrote Dr. Stanley. The "gates" to which he referred were in reality merely strands of wire strung across short vertical poles and stretched between two fenceposts. On one professional call, Dr. Stanley remembered, he struggled with nineteen of those gates each way.

Notre Dame Street, Montreal. An undated lithograph from a drawing by James Duncan (1805–1881).

9

Developments in Street Paving

IN THE GREAT CITIES OF THE EAST, the streets in the last decades of the nineteenth century had come a long way from the mud-tracks of pioneer days. The main thoroughfares were paved. Over them gentlemen in top hats and ladies in full crinolines rode smoothly in open victorias, probably the most graceful carriage ever invented. Horse-drawn streetcars rattled along on rails. Heavy wagons, loaded with kegs and sacks, rumbled behind teams of matched draft horses. Traffic jams were still fairly rare, but a steady flow of carriages, cabs, and delivery carts filled the principal roadways. Transportation within city boundaries had reached a relatively high level of comfort and efficiency to meet the requirements of urban society and commerce.

Improvement had come gradually and in stages. Details varied from city to city: on Quebec City's steep hills, for example, cobblestones and bricks were used far more widely than on the more level thoroughfares of Upper Canada. We may trace the general development, however, in a single history—that of Toronto.

In the city's earliest days the major problem, once the streets had been cleared of trees, was the stumps the axemen left behind. They were a nuisance to avoid by day, and a hazard by night. In at least one instance roadbuilders tried to fell the trees near ground level and smooth the stumps off with the adze; but the stumps sprouted and it became clear that the only permanent solution was to eradicate them entirely. About 1800, public drunks were commonly sentenced to a period of removing stumps from the roadways.

Pathmasters and overseers had to battle encroachments by landowners who put up buildings or fences inside the statutory road allowance of sixty-six feet. In 1811 eleven persons were ordered to remove such obstructions. The open drains also gave trouble. In the dark it was easy to stumble into them unless you carried a lantern. Then, too, they became clogged. In 1816, as part of a programme to improve King Street, the town's principal commercial road, the pathmaster was instructed to "open water-courses on each side which have been arrested in their natural course either by negligence, inattention, or indifference, not to say worse." Footpaths on either side of King Street were to be eleven feet wide; merchants who wanted to provide a raised sidewalk to protect their customers from mud and dirt had first to seek permission of the magistrates. The roadway itself was to be built with a crown so as to drain properly.

Before the War of 1812 some attempt was made to pave footpaths with flagstones from the lakeshore, but this was done only at a few important locations. It is said that the stones were irregular in size and crudely set, so that the results were far from satisfactory. Yet it was probably better than walking through dust or mud. For such projects statute labour was employed as well as public funds.

Streets were unmarked by name until 1830, nor were there any numbers on the houses. In that year the Clerk of the Peace was directed to "take steps to get the names of the streets painted and put up wherever there is a corner house." Two years later a small appropriation was made for numbering dwellings.

York was incorporated as the city of Toronto in 1834. As one of its first actions the new council increased the tax rate and borrowed money to lay wooden sidewalks. These consisted of two planks running side by side on stringers. To save money they ran only on the side of each street which had the heavier pedestrian traffic. About eight miles of sidewalk were built. Jesse Ketchum, a well-known tanner whose business and home were too far (a few hundred yards) from the city centre to be so served, kept the footpaths in his area in good condition by scattering a thick coating of tanbark.

By 1834 upwards of two miles of King Street had been flagged and paved with stone. About the same time it was suggested that crosswalks at the main intersections should all be paved with stone to save boots and skirt hems from the muddy roads.

"Macadamizing" began on King and Yonge Streets in 1836. If McAdam had seen the work he would have had it torn up and started all over again. There was no base of boulders, the road was laid on only about nine inches of stone—and that roughly broken, for the cost of breaking it was almost as much as the stone cost, and about double the expense of laying it. Repairs were required constantly. To make matters worse a common sewer was placed under the centre of King

St. Urbain Street, Montreal: for many years the best surface on city streets consisted of well-packed snow. Yet it was possible to overestimate the efficiency of runners on snow. The cartoon at right, from the Canadian Illustrated News of 1871, was labelled "Cruelty to Animals."

In 1870 the steep grade of Abraham Hill, Quebec City, was unpaved, though pedestrians could follow a narrow board sidewalk.

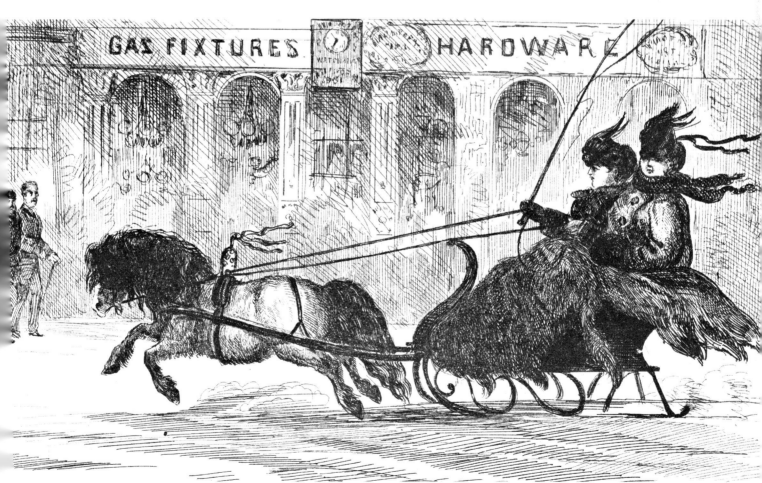

Street, and the earth above it was not properly consolidated; this, more-over, was the beginning of a new breed of traffic obstacle, for the road thereafter frequently had to be torn up so that new feeder drains could be installed from the sides.

But the age of plank roads was approaching. When the new technique was still only an experiment on the highway, James Beaven, author of *Recreations of a Long Vacation*, found that in Toronto

considerable portions of some of the principal streets have had their carriage ways planked under the direction of the corporation. In these cases the plank is made to slope from the middle to the sides, and where the work is completed the space of about ten feet on each side, between the carriage way and the footway, is paved with boulders.

James Buckingham did not remark on this improvement when he visited in 1839, but he was impressed with the general state of the city's thoroughfares:

The centres of all the streets are either paved or macadamized; and the side-walks are chiefly, though not entirely, of wooden plank, placed longitudinally as on a ship's deck, and forming a far more clean, dry, elastic, and comfortable material for walking on than any pavement of stone or brick. In the few instances, indeed, in which flat stone-pavement is used instead of wood, it is extremely disagreeable to pass from the latter to the former.

The new city's first mayor, William Lyon Mackenzie, is credited with most of the improvements in the sidewalks, and with providing wooden crossings, sometimes longitudinal, sometimes transverse, at street corners. The planks were fastened with wooden plugs or with long nails, and were chopped off at the edges to facilitate carriage travel, but it was difficult—as with the earlier flagstone walks and crossings—to keep them free from bumps and holes.

When the horse railway was inaugurated in 1861 a better macadam roadbed was laid along its route on Yonge Street, on Queen west of Yonge, and on King from Yonge to St. Lawrence Market. The railway company was to maintain the road between the tracks and for eighteen inches on either side; the city saw to the rest. On other streets "maca-damizing" consisted largely of dumping stone in the centre of mud roads—the common practice everywhere—from which it was kicked about by such traffic as did not avoid it, providing excellent ammuni-tion for riots during elections and Orange Lodge or Roman Catholic processions.

The street railway company introduced cedar-block paving to Toronto in the late sixties. The blocks, cut six inches long from poles, were laid end-up on sand between the ties on which the rails were laid. They gave a hard, long-wearing surface to withstand the constant pas-sage of the horses pulling the streetcars. During the next decade the city paved the entire width of some of its most heavily travelled streets with cedar, pine, or spruce blocks. In those days the roadway was bordered with deep gutters. Board sidewalks were raised high above street level to escape dirt thrown up by wheels, and pedestrians crossed over the gutters on bridges.

In the same period some streets were paved in brick, which was particularly effective on hilly sections. In 1885 a traveller, Thomas

The horse car introduced improved paving to Toronto. Because it ran on rails, the horses were confined to a relatively narrow path which quickly became badly worn. To counter the effects of their hoofs the company first laid cedar blocks, and later bricks, on the right of way.

Street maintenance in Prince Albert, Saskatchewan, in the first decade of this century. A simple drag levelled out the worst of the bumps.

Each new town began in the same way. This is King Street in Golden City, Ontario, during the Porcupine gold rush of 1911.

Great St. James Street, West Montreal, was paved by 1870.

Richman, reported that the streets he passed along were paved with "coloured boulders of syenite, gneiss, limestone, porphyries, and granite."

Though portland cement was first manufactured in 1824, it did not influence Toronto streets until the late eighties. The first "granolithic" sidewalks were laid in 1886, in front of the Rossin House (later called the Prince George) and at the corner of Yonge and Front Streets. They consisted of a six-inch bed of concrete with a wearing surface of cement and granite chips.

Asphalt was first applied to street surfaces in 1888. The following year the street railway, soon to be electrified, began using concrete under its ties. In 1893 it introduced brick paving between the rails. As the years passed, sidewalks and paved streets became more and more common. But it was not until 1904 that the cedar blocks were torn up along Yonge Street and replaced by asphalt.

Main Street, Winnipeg, about 1882, looking south. Prairie streets were built wide so that drivers could manoeuvre easily around ruts and potholes.

Government Street, Victoria, in 1871—a broad expanse of wheel tracks bounded by raised boardwalks.

Winter once was a boon to the traveller. The streams and rivers froze; snow smoothed out the bumps and, packed down, provided a good surface for sleighs of all shapes and sizes. The farmer and his family, bundled up for warmth, drove over the fields and waterways to visit friends. Residents of the cities and garrisons made more elegant jaunts— a favourite for those in the Quebec area being the sleigh ride to Montmorency Falls.

Trotting races were held on the ice near Montreal in the 1870s. In Toronto amateur sportsmen raced their sleighs down the broad expanse of fashionable Jarvis Street, to the delight of neighbouring children.

But for the first horse through deep snow the way was far from easy. The phlegmatic acceptance depicted in the old print was mirrored in a photograph taken in this century near Kirkland Lake. In such snow horses could last only a few hundred yards before they required a rest.

With the rise of cities and the progress in transportation, winter lost most of its advantages. The horse-drawn streetcars ran on tracks which had to be kept clear of snow. This situation led to a battle of legendary proportions in Toronto during the winter of 1880–81. The Street Railway Company hired men to shovel snow from the road onto the sidewalk, blocking the entrances to the Yonge Street stores. In retaliation shopkeepers and their clerks joined in and shovelled the snow back onto the tracks. By the time they were finished, public transportation along the city's main street had been forced to a halt.

WINTER

THE FIGHT BETWEEN THE STOREKEEPERS AND THE COMPANY'S EMPLOYEES.

THE BLOCKADED CARS AFTER THE BATTLE.

TORONTO.—SNOWING UP THE TORONTO STREET CAR COMPANY.—From Sketches by W. N. Langton.—See Page 99.

"A street nuisance of the worst kind," the Canadian Illustrated News remarked of the snow and slush in the winter of 1871. "A much-enduring public has to take refuge in thick boots and rubbers. But even when armed with these protections walking at this time of the year in the crowded streets is rather an undesirable exercise, and one calculated to try both temper and shoe-leather."

In the smaller towns and in the country, sleighs were still in common use within the author's memory.

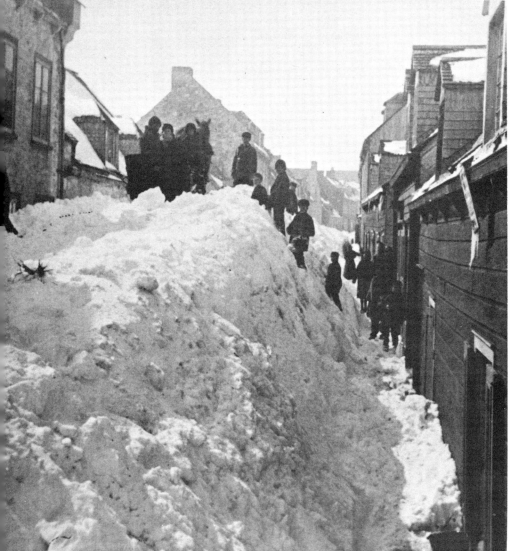

This street scene in Quebec City, about 1870, is an extreme example of a common practice. Houseowners cleared the snow from in front of their doors and piled it in the middle of the road. Traffic, including horses, followed the highroad—no matter how high it grew.

In 1931 some Model T Fords were equipped with skis and tracks to travel on uncleared routes. In Toronto by that time the Transportation Commission was using snow ploughs (bottom centre) to keep its tracks clean without controversy. Today winter is no longer the friend it once was, but neither is it the formidable adversary it became for a while. Road-builders have added new weapons to meet its extremes. The plough is still the traditional standby, but now it is supplemented by snow blowers (bottom right), and by chemicals and abrasives which cut through ice and snow to provide traction. The cost is enormous— upwards of nine million dollars in some years for the streets of Montreal alone, plus the cost in chemical corrosion to vehicles and highway structures, for no economical substitute for common salt on the roads has been found. Yet in the 1960s snow clearance has become so effective a team operation that a blocked highway is news, and Canadians drive as a matter of course in every season.

10

The Railway and the Buggy

WHILE THE STREETS OF THE MAIN CITIES of eastern Canada—and of Vancouver in the west—were improved to meet the demands of concentrated metropolitan traffic, the highways fell into neglect. In the twentieth century they were to spurt ahead to undreamed-of grandeur, but first they had to pass through their blackest period. The second half of the nineteenth century belonged to the railway. As the gleaming steel rails snaked smoothly across the countryside, roads were abandoned by all but local travellers. Expenditures were cut to the minimum necessary to keep them in repair—and often far below that. As the stagecoaches disappeared hundreds of miles of trunk road were allowed to deteriorate from the small degree of excellence so painfully achieved.

The first short stretches of railway, like the first roads, followed the old portage routes. In 1836 rails were laid around the Richelieu rapids to speed travel between Montreal and New York; three years later around Niagara Falls; in 1847 along the Lachine rapids above Montreal. And in the next decade large-scale building began in earnest. During the eighteen-fifties and early sixties more than two thousand miles of track were laid down between the major centres of population. For the most part the new lines paralleled existing roads—from Halifax to Windsor and Truro; from Saint John to Moncton and Shediac; from Quebec eastward to Rivière du Loup; from Quebec westward to Montreal and on to Kingston, Toronto, Hamilton, London, and Windsor; from Toronto to Sarnia; from Toronto northward roughly along the Yonge Street route to Lake Simcoe and Collingwood. Smaller lines sprang up to feed the trunk routes, particularly in Canada West (as Ontario was then known) where every municipality and county sought service from the iron horse.

Almost overnight long-distance transportation ceased to be an endurance contest. Granted that the early railways were slow by modern standards (top speed 25 to 30 miles an hour, and an average speed of somewhat less than half that); that the sparks and ashes flying from full-bellied smokestacks were a nuisance to passengers and a danger to forests; that without snow ploughs trains could be trapped in drifts for hours or even days during storms: nevertheless, for the first time Canadians had a convenient means of travel in all but the worst winter storms, behind a tireless steed which fed on the wood in which the country abounded. The train could haul heavy loads when the waterways were frozen; it could carry passengers when the spring roads were impassable; it rode smoothly on flanged wheels over steel rails, without fear of mudholes or the jolting of ruts and corduroy. After 1870 it was even possible to sleep comfortably while being carried across the countryside in the new Pullman cars. No wonder that trains took over most of the traffic along their routes, or that the roads were left to the horse and buggy and the farmer's wagon.

But railways were expensive. By Confederation nearly $150 million had been spent on their construction. Much of the money came from private investors in Canada and England; much also came, however, from the limited provincial treasuries, and from municipalities which borrowed beyond their means to ensure that the railway would not pass them by. Even so small a centre as Port Hope, with about

In the late nineties horse-drawn vehicles of every description streamed along the main streets of the bigger cities. (Drawing by Leslie Saalburg.)

five thousand people, raised a loan of $740,000 to invest in the new form of transportation. There wasn't much money left for roads, and what there was was watched carefully. In 1892 the weekly newspaper in Shelburne, Nova Scotia, cast a critical glance at the use made of a $30 grant for work on a local road and bridge. A call for tenders had resulted in a bid to supply three thousand feet of plank for the bridge at $7.75 per thousand feet, or $23.25 in all. The newspaper estimated that the lumber could be put in position for another seventy-five cents. With a reasonable commission this brought the total expense to $25.50. "There are some," the editor continued sternly, "who are much interested in the said road who would like to ascertain from the Commissioner what has become of the balance of $4.50."

The system of statute labour was still in force, and should have seen to maintenance and improvement of the roads. Human nature being what it is, however, few landowners were prepared to expend much effort or money to this end—even on the routes in front of their own property. John C. Geikie has told of a typical day of statute labour in a book published during the railway era.

My three elder brothers and a number of neighbours were on the ground on the day appointed, but they were an hour or two later than they would have required any labourers they might have hired to have been, and they forthwith commenced their task. It was amusing to see how they managed to get through the time, what with smoking, discussing what was to be done, stopping to chat, sitting down to rest, and all the manoeuvres of unwilling workers.

The roads were kept usable, but that was all. Many trunk roads grew worse with the passage of time. Even the turnpikes fell into disrepair

The ubiquitous buggy, most popular of horse-drawn vehicles, on the road past Port-au-Port, Newfoundland.

as revenues dwindled in the face of competition from the railways.

Throughout most of the second half of the nineteenth century, construction of new highways seems to have come to a halt in eastern Canada. Here and there undoubtedly local wagon trails were opened and gradually improved as the population of a given area increased, but these only supplemented the existing network. No major new roads appear to have been constructed in the Maritimes or Quebec. Ontario still required new colonization routes as pioneers continued to press northward into the bush, but the roads opened during the early sixties in the Muskoka and Kawartha regions were no better than those of a half century before. While millions were spent on railways, these rough trails were pushed through the rocky wilderness at a cost of $250 to $400 per mile under primitive, often oppressive, conditions. Labourers, desperate for work, were obtained from neighbouring settlements at wages as low as fifty-seven cents a day, when nearby farmhands were being paid one dollar; and many a contractor added to his profit by providing low-grade food. There was little sympathy for the workers. One inspector, on hearing of an attempted strike, wrote to the Crown Land Agent in charge that every man should have been discharged forthwith.

I am a little surprised [he continued] that you should have submitted to parley a moment with men engaged under such circumstances. We cannot submit an instant that those who are hired under charitable considerations should dictate to us the terms of contract. . . . There is just one simple method of action—*discharge the dissatisfied*—dont keep grumblers on the work, they'll infect the good. Act firm and prompt in this matter or your men will get the mastery over you.

The roads that resulted were as rough as their builders, as anyone knows who has travelled over them, and many of them persisted, with or without improvement, down to modern times. They wound back

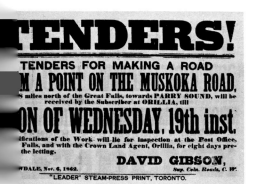

Buggies and tradesmen's carts crowded the Belleville, Ontario, marketplace.

and forth, around rocks and over the granite ridges of the Canadian Shield. Many were passable only by oxen or horses. Where lumberjacks were at work the roads might be destroyed in a single season by the heavy loads of food and supplies needed to maintain a camp. Elsewhere nature threw up obstacles of her own: underbrush speedily returned in luxuriant profusion, and in summer some soils produced rank growths of timothy through which low-slung wagons had to literally slice their way. In rocky areas floods washed out many a bridge and causeway in the first spring after it was built. Those that survived the floods were apt to be destroyed by forest fires: on one road in one year, nine bridges went up in flames in the first twenty-three miles. For want of adequate roads many settlers simply gave up their farms, and those who stayed in some cases paid as much as $1.25 to have a hundred pounds of supplies carted less than thirty miles. In central Ontario in this period, bad roads hampered settlement as much as bad land.

West of the Great Lakes there were no railways until the transcontinental line of the Canadian Pacific was built in the 1880s. Immediately, its steel ribbon outshone the dusty wagon trails. No form of transportation dependent on muscle-power could match the iron horse over the vast distances of the prairies and the mountain slopes. Roads became little more than feeder routes, on which immigrants and manufactured goods travelled from railway stations to the scattered farms and produce was carried back to the trains to be hauled east.

Even the mighty Cariboo Trail suffered indignity, for much of its path through the Fraser Canyon was destroyed in the building of the CPR. On its northern stretches, however, the road to Cariboo was as busy as ever—though now it served ranchers instead of gold miners. Supplies for the ranches were shipped by rail as far as Ashcroft; from there, on the route past Clinton to the north, the old freight wagons and stagecoaches took over. Veterans of those years still tell of a hundred outfits spending the night at one stopping place.

This is a good example of a relationship which held true across the country. The roads were neglected, but never unused. The railway took the lion's share of money, glamour, and long-distance traffic, but it needed the roads to survive. Without them it could not serve the communities and settlements which lay on either side of its right-of-way. To a limited extent the coming of the railway even encouraged road-building. This was particularly true in Newfoundland, where the rocky interior had successfully withstood penetration for centuries. Once a twisting narrow-gauge railway had been forced across the island from St. John's to Port-aux-Basques in the early nineties, wagon trails began to appear connecting the seacoast villages with the tracks.

Even in Ontario and Quebec, where railway construction was heaviest, the roads never lost their importance for local traffic. The farmer still drove to market, and hauled his own grain to the nearest mill or distillery. As dairy farming developed, the rough roads also were used to carry milk to the local cheese factory. Sometimes an enterprising farmer made a business of collecting his neighbours' milk and carrying it for them to the factory for a small fee. We have a record of one such

Buckboards and blacksmith shops were still a common sight in 1910.

arrangement in 1895: for hauling the milk cans six miles to the cheese factory and back again every weekday for six months, and every second day in the seventh month, this particular farmer earned $150—not much by modern standards but a handy amount when cash was scarce. The economic historian Harold A. Innis has described how his interest in transportation began in part with memories of drawing milk to cheese and "condense" factories as a boy in western Ontario.

Butter and eggs also travelled by road from the farm to the general store, where they were exchanged for groceries and dry goods. There was not always a good sale for these dairy products, for many towns-people in those days kept a cow and hens; crocks full of rotten butter and eggs frequently had to be thrown out by the merchants. In the seventies, however, preserving techniques began to be developed for storing butter and eggs, and in the late seventies and early eighties the first butter factories or "creameries" with ice refrigeration were established. Merchants then sent wagons to the farms to collect produce and to return a day or two later with groceries and supplies in payment; as a boy the writer went on such trips in the Cobourg–Rice Lake district of Ontario. Some merchants also collected surplus butter from the general stores. Among the latter was young Joseph Flavelle, who had a creamery in Lindsay, Ontario. Years later, after he had become a millionaire meat packer and a baronet, Sir Joseph recalled, "For weeks in the fall months, I drove to villages and towns, buying or endeavour-ing to buy, the store-packed as well as the dairy-packed butter, which . . . in due course was exported to London or Liverpool."

The roads were still the normal means of going to church and for social visiting in the country. The most common passenger vehicle—amidst all the varieties of carriages and coaches, chariots, traps, victorias, phaetons, hacks, broughams, and sulkies—was undoubtedly the buggy.

Collecting milk for the cheese factory at Vernon River, P.E.I.

The general store and post office were the goals of many a rural drive.

A good team and an elegant carriage were everyone's dream.

Smart turnouts were a matter of pride when this Toronto delivery wagon was photographed in 1884. Horse shows gave prizes in many commercial classes.

These high-bodied open carriages gave the period its name: the horse-and-buggy age. In the 1890s, when carriage production reached its peak, scarcely a settlement of any size in eastern Canada was without a factory making them. Buggies were fitted with long steel springs, hickory wheels, a collapsible top, and a fancy dashboard; but what made them particularly popular among the younger generation was the single seat. A summer evening's ride in one was a recognized stage in rural courtship. Late in the century the solid rubber tire, developed in England in 1876, began to catch on; people were wary at first of this improvement, for fear that "silent" vehicles would be dangerous, but most were converted eventually by such advertisements as this:

No jolt. No jar. All vibration, which is so spine-racking and debilitating to nervous people, is taken away. They make the vehicle run absolutely noiselessly, permitting conversation in a low tone of voice; make the roughest roads smooth; prevent slewing on street car and railroad tracks.

Of course not even rubber tires could do very much to soften the passage over the bumps, holes, ruts, and washouts which were still normal hazards in the Gay Nineties. But help was on its way. Two new inventions had appeared which would revolutionize the nature of transportation. For Canada's roads a new day was about to begin.

Many bridges were covered to preserve the wooden decks. The 155-foot span over the Annapolis River, Nova Scotia (above), was replaced with steel in 1907. The bridge at Napanee, Ontario (below), lasted from 1840 to 1909.

Above: Indians built the first Haguevelget Suspension Bridge using suspension cables of copper wire left from a telegraph line. British Columbia engineers later built a gorge-top span. Below: Timber bridge on the Matapedia Road, Quebec.

A railway suspension bridge was built across the Niagara Gorge about a mile and a half below the Falls in 1855. Vehicles and pedestrians crossed on a semi-enclosed lower level (photographed at left) some 230 feet above the river surface.

Tabusintac River Mouth Bridge, New Brunswick, built in 1897, was dwarfed by its approaches, which were of cedar open cribwork to let the tide flow freely. The bridge proper consisted of two timber open truss spans and a timber swing span.

BRIDGES II

The world's longest cantilever bridge crosses the St. Lawrence six miles above Quebec City. It was completed in 1917 after twice collapsing during construction, killing eighty-eight men. The main span is 1,800 feet long.

This suspension bridge over the St. John River was completed in 1860 after an earlier bridge on the same site collapsed in thirty-below weather.

*Over roads built for buggies advanced
the pioneers of the horseless carriage.
But it was a rare journey which did
not require at least one stop to pump
up a tire.*

11

Advent of the Automobile

FIRST CAME THE BICYCLE. The first in Canada was imported from England in 1878 by John Moodie of Hamilton when he was eighteen years old. It was a strange-looking contraption, with a front wheel five feet high and a back wheel only about a foot in diameter. It was dangerous, too, for the rider perched on a seat several feet above the ground. If he lost his balance it meant a long fall entangled with a heavy machine; going downhill, without brakes, there was the added risk of hitting a stone or bump and sailing headfirst ("taking a header," it was called) over the handlebars. For two years Moodie had the only bicycle in Canada. Then a Toronto man, Harry Goulding, imported five more, and in 1881 the first Canadian bicycle race was held at the Toronto Industrial Exhibition. Moodie came in second in a field of three. A year later the Canadian Wheelmen's Association was formed, and cycling grew increasingly popular.

It was hardly a comfortable sport as yet. Big wheels, solid tires, and rough roads earned these early machines the name of "boneshakers." The "safety" bicycle, when it appeared in the late eighties, was a great improvement. Its wheels were both approximately the same size, and before long were fitted with pneumatic tires. Its tubular metal frame was light and easy to manage. It was powered by a chain-and-sprocket drive arranged to make the back wheel turn faster than the pedals. Best of all, it was cheap. The clumsy old "boneshaker" was a rich man's toy, costing anywhere up to $300. The "safety" cost as little as $25.

It is hard for us to realize just what the bicycle meant in the nineties. Here was everyone's dream—a speedy, pleasant means of travel available to people who could never afford a carriage and team of horses. Overnight Canada became a nation on wheels. Bicycle clubs sprang up in every town and city. The younger generation pedalled out to the country on picnics; families got out a bicycle-built-for-two, with extra carriers fore and aft for the youngsters, and pedalled to the beach. The man who wanted to prove his strength cycled a "century," one hundred miles, in a single day—no mean feat over the roads of the times. Others raced in bright club sweaters. The "scorcher," or speeder, became a public menace. After the drop frame was introduced to preserve feminine modesty, many women took up the sport. Most wore knickers or long, baggy bloomers under their long skirts, but even that wasn't enough to calm the critics, and a few women were arrested for wanton exposure of their ankles. Nevertheless the bicycle, by giving women a new mobility—and an individual and unchaperoned mobility at that—did much to speed their eventual emancipation.

And after the bicycle came the automobile. Canada's first was an electric car built in Toronto in 1893 by J. B. Featherstonhaugh. It looked like an oversized baby carriage. The first gasoline-powered car in the country was owned by the same John Moodie who had the first bicycle. It was a one-cylinder Winton, bought in 1898. Like most automobiles of the time, it still closely resembled a horse-drawn carriage. The motor was under the seat, so that in front there was nothing but a low leather dash; the wheels, thirty-six inches high and fitted with three-inch tires, could have come off a buggy. Moodie had quite a

time getting the vehicle across the border from the United States. No such thing as an automobile appeared on the tariff schedule. The customs officers considered it a carriage, and wanted to charge thirty-five per cent duty; its owner argued that it was a locomotive, which bore only a twenty-five per cent tariff, and won his point. This was not the end of his troubles, however. That summer Moodie drove his car to a public park, where a camp meeting was being held. The Winton attracted so much attention that he was finally asked to leave so that the meeting could continue. He refused until his entrance fee was returned; then he threw off the muffler and retired slowly, with all the noise and splutter that an early car could make.

By that time the gasoline buggy was fifteen years old, but so few had been produced that every one was a novelty. The first had been manufactured by Karl Benz in Germany in 1885, and with Gottlieb Daimler as a partner Benz went on to build some of the world's finest cars. The earliest American-built machine with an internal combustion engine was produced by the Duryea brothers, Charles and Frank, in 1893. Six years later thirty small companies were manufacturing cars in the United States, but the total annual output amounted to only six hundred vehicles. Some of the early types had wide "baloney" tires with steel rims, brakes which often could not hold the car halfway up a hill while the gears were being changed, and small metal-and-glass boxes in which candles were lit before an evening drive.

The gasoline buggy was still a long way from perfection, and people were not yet convinced of its merit. Most of them preferred to travel

Canada's first gasoline-driven automobile, a one-cylinder Winton, with its proud owner, John Moodie, at the "cow's tail" steering handle.

behind a team of spirited horses. Others favoured the steam-driven automobile. There were several makes of these, all burning kerosene to heat water in a boiler to make the steam which made the wheels turn. Some "flying teakettles" were remarkably efficient: the 1904 Stanley Steamer, with throttle wide open, could reach forty miles an hour. There were also electric cars which moved silently and smoothly on the power of thirty to forty storage batteries. For some years all three forms of locomotion competed, with some doubt about which would win. The steamers, however, were complicated to keep in tune, somewhat dangerous, and took half an hour to build up a head of steam. The electric motors started as soon as the current was switched on (an advantage to doctors and others who had to make many stops in the course of a day), but after a hundred miles or less their batteries had to be taken to the garage for recharging. In the end, of course, the gasoline engine triumphed.

By 1901 the automobile had ceased to be an oddity. That year, mass production was initiated by Ransom E. Olds. His famous runabout still looked like a carriage, with its curved low dash, spoked wheels, collapsible top, and brass side-lanterns, but its price was only $650. In four years the factory turned out twelve thousand cars, while the world sang "In My Merry Oldsmobile." In 1908 Henry Ford began producing the Model T on an assembly line, and placed the car within the reach of everyone. His famous "Tin Lizzie," stripped down to the absolute essentials, sold for as little as $290 for the roadster model. Before the Model T was discontinued in 1928, he had made fifteen million of them.

Canada's motor industry had its beginnings in 1904 when a group

Mass production placed the automobile within popular reach and changed Canada's travelling habits. This early assembly line is at the Russell Motor Company plant in Toronto.

A stylish turnout of 1910. Inspired by such a scene, Percy Punshon, an occasional verse writer who lived close to the new auto factories of Oshawa, Ontario, penned an ardent tribute to progress. It ended:

*When I think of the motor-car silent
    and swift,
In the dark sky of transport, so long
    overcast,
I see in my vision a limitless rift
That heralds the dawn of the best
    thing at last.
There is comfort and speed in the
    automobile
That none can deny, and no one
    gainsay;
Then I know that with me you will
    readily feel
How thankful I am that I'm living
    today.*

An early, chain-driven truck.

of Windsor businessmen organized the Ford Motor Company to manufacture and sell Ford's early products in Canada. Little more was involved than assembling bodies and wheels on chassis ferried across the river from Detroit. There were then about six hundred cars on Canadian roads. In 1907 R. S. McLaughlin designed and built an automobile in the McLaughlin Carriage Company plant at Oshawa, Ontario. Because of the high cost of producing engines and some other parts in small numbers, he arranged to buy them from the Buick plant in the United States, beginning a partnership which would eventually bring him to the chairmanship of General Motors of Canada. In that year, 1907, 2,131 cars were registered in Canada, most of them in Ontario, Quebec, and British Columbia. After that the number skyrocketed. By 1912 Canada had fifty thousand cars. In 1917, the registration of cars and trucks approached two hundred thousand—roughly one for every forty persons in the country.

No one could imagine when it began how rapidly the motor industry would develop. In 1904 one authority predicted optimistically that within forty years the United States would have sixty thousand automobiles; before he was five years older the annual production of new cars alone had reached twice that number. The result was a revolution, not only in transportation but in the entire way of life.

At the turn of the century a journey of even two hundred miles was for most people an extraordinary occurrence—something that happened only once or twice in a lifetime. It was usually by train, and according to a schedule and route set by the railway company for its own convenience. The bulk of the population lived in the country. For them even a short trip to town was an event and a chore. The horses had to be led out of their stalls, hitched to the carriage, driven to town and back, then watered, fed, and brushed. If they got sick they had to be cared for. The automobile in comparison was handy and economical, as Gottlieb Daimler pointed out in an early advertising jingle:

> He never eats while in his stall,
> Drinks only when he starts to haul,
> Plays no dumb tricks that vex or tease,
> Contracts no hoof and mouth disease,
> Raises no angry heel to kick,
> Nor raids your standing grain, or rick—
> So buy this beast of happy ways
> And live in clover all your days!

Further benefits were foretold by the editor of *Horseless Age*, when that magazine made its debut in 1895. "In cities and towns the noise and clatter of the streets will be reduced, a priceless boon to the tired nerves of this overwrought generation. . . . Streets will be cleaner, jams and blockades less likely to occur, and accidents less frequent, for the horse is not so manageable as a mechanical vehicle." Those particular predictions have not stood the test of time very well. In one matter, however, the editor was correct: the horseless carriage did save enormous amounts of time and money. It gave the world a freedom of movement unknown to any other age. The car provided fast, cheap, easy access from the farm to the city, from the city to the country, from home to work, to shops, to schools, to recreation. For the first time a private

This speed demon was going 15 miles per hour when the traffic policeman overtook him—on a bicycle. Forced to the curb, the unhappy motorist receives his summons. The spectacle drew quite a crowd, in Montreal back in 1913.

means existed, available to the individual and his family, to annihilate the great distances of the North American continent.

Yet recognition was slow in coming, and for the early motorist the way was far from smooth. The new invention—admittedly noisy, smoky, and frightening—provoked an animosity and opposition which in retrospect seem incredible. Prince Edward Island totally banned its use outside the cities. In Ontario the law required every motorist meeting or passing a horse-drawn vehicle on a country road to come to a full stop; if the horse displayed alarm, the motorist had to climb down and lead the animal past his vehicle. The speed laws were also restrictive. Moodie found a limit in Hamilton of seven miles per hour. The speed trap seems to have arrived with the automobile, for as early as 1904 *The Motor* published a parody:

> Screened by the wayside chestnut tree
> The Village P.C. stands.
> The "cop" a crafty man is he
> With a stop watch in his hands.
>
> . . . .
>
> He goes each morning to his lair
> And hides among the trees,
> He hears the sound of motors there
> And it sets his mind at ease
> For it seems to tell of captives—and
> Promotion follows these!

In some parts of Canada traps of another kind awaited. Many farmers protested that automobiles frightened the livestock, and to discourage motorists they buried planks studded with spikes, points uppermost, in the roadbed. Other farmers took advantage of the interloper to turn a dishonest dollar. A favourite trick was to prepare a mudhole and charge five dollars to haul each unwary motorist out of it. A. C. Emmett, for many years managing director of the Manitoba Motor

League, used to patrol the roads with other Winnipeg drivers in search of such malefactors. Early one morning they caught a farmer just preparing for the day's business. In his wagon stood two big barrels of water to soak the hole, and a pile of straw to scatter over it. Summary justice was dealt out. They dipped the culprit in his own barrels, and forced him to draw gravel to fill the hole.

In other parts of the country it was scarcely necessary to improve on nature. The roads were still the same ones which had been opened for horse-drawn wagons and oxcarts. Here and there were a few "macadamized" stretches, but most were common dirt roads, unsurfaced, ungraded, full of holes and ruts, dusty in fair weather, and mud traps after a rain. Before setting out in an open car travellers bundled up in full-length coats, or "dusters"; the women donned wide-brimmed

hats and veils, and the men wore special caps and goggles. En route they had to share the way with cattle, horses, sheep, and goats, which enjoyed equal freedom of the road. Service stations were few and far between: every motorist was his own mechanic. He had to be his own pathfinder, too, for road maps were a thing of the future. At best he had a routing card bearing such instructions as: "Mileage 3.6. Watch road for bad culvert. Mileage 6.2. Turn left with red barn on right." There was always a chance, of course, that in the interval since the instructions were printed the barn had been painted another colour. A trip of one hundred miles was an achievement to boast of for months afterward.

J. A. McNeil, one of the best-known pioneers of Canadian motoring and for years general manager of the Canadian Press, has described a

Guiding a 1912 White over a difficult stretch.

Cross-country driving once required considerable navigational skill. Directional signs were far from common. In several provinces highways were identified at best by coloured bands painted on the telephone poles, so that motorists could safely follow the yellow (or blue or red) road to another city. Highway numbering is said to have been proposed in 1920 by A. C. Emmett of the Manitoba Motor League; it quickly became the standard method throughout North America.

journey from Toronto to Walkerton in September 1905. It took eleven hours, plus an overnight stop, to travel the 120 miles.

Leaving Toronto at four P.M., we reached Guelph 50 miles and four hours later without incident, except for stopping each time we met a horse-drawn vehicle, and when necessary leading a fractious steed or team past our car.

Spending the night in a Guelph hotel, we resumed our trip about eight A.M. Approaching the village of Harriston in mid-morning, we had a puncture and stopped a few hundred feet from a country school while our driver repaired the tire. The motor car was so new in that long-settled section of Ontario that the school teacher allowed her pupils a premature recess to view the strange vehicle. We were surrounded by a group of curious, chattering children.

Neighboring rustics gathered, too. One was a grizzled patriarch who told us he was walking to the village two or three miles on. Our kindly driver offered him a lift when the repair was finished. Raising both arms in a gesture of horror, the old fellow ejaculated: "Me ride in that thing! No! No!"

Resuming our run, we encountered on the narrow road another of the hazards of pioneer motoring—a large herd of browsing cows (which like horses, sheep and pigs were then allowed at large on the highways).

Stopping until they had leisurely cleared the roadway, our driver edged cautiously forward but not slowly enough to avoid striking a calf which, separated from its mother, responded to the alarmed maternal "Moo-o-o-o" by dashing to rejoin its parent.

I made the return trip to Toronto by train in a little more than five hours.

McNeil and his fellow drivers were adventurers. To them, much of the appeal of driving was to triumph over all obstacles. But as the number of automobiles grew, the need for better conditions became

increasingly clear. Horse-and-buggy roads were not good enough for the new mechanical mobility. To improve them would require a great deal of money. Moreover, as long-distance travel returned to the highways it became obvious that their care and development could not be left to municipal governments, statute labour, or private enterprise. The automobile demanded planning, construction and maintenance of roads on a grander scale. The provinces would have to play a larger role, and so eventually would the federal government.

At long last, the people associated with roads began to organize. The movement commenced during the cycling craze, for poor roads inconvenienced the cyclist as well as the motorist. In 1894 the Ontario Good Roads Association was formed to disseminate information concerning highway construction, and to inform the people of the economic importance of a good road system. It was followed by similar organizations in Quebec, Manitoba, and other provinces. In 1914 the Canadian Good Roads Association was created to press for highway improvement at all levels of government. By that time the success of the provincial organizations was already evident.

Forerunner of the modern house trailer was this "Pullman automobile," designed in 1911 by nineteen-year-old Henri Dandurand, son of the Montreal automotive pioneer. Built specially for the large Dandurand family, which included eleven children, and their friends, it could carry twenty-six persons seated or twelve sleeping. A pullman-type body was set on a three-ton Packard truck with an extremely long wheel base. The entire vehicle was twenty-nine feet, three inches long, but only six feet, three inches wide. It never left Montreal Island because of its weight (11,700 lbs.): none of the existing bridges could support it.

*The interior of the Dandurand trailer
was luxurious even by today's
standards. There were eleven stained-
glass windows, twenty-one electric
lights, and phones between every
compartment. The machine cruised
on solid rubber tires at twenty-eight
miles per hour. It had a set of
hydraulic shock absorbers so that it
rode with the motion of a ship.*

12

A New Day for Roads

ONTARIO WAS FIRST to respond to the cry for new highways. In 1896, at the recommendation of the fledgling Good Roads Association, it appointed a provincial "Instructor in Roadmaking." Because there was not yet a separate Department of Highways he was attached to the Department of Agriculture on the grounds, presumably, that his task was properly part of improving the countryside. Alexander W. Campbell had earned an enviable reputation as city engineer of St. Thomas, Ontario; in his new position he urged the need for progress with extraordinary power through the popular newspapers and his own inspired official reports. In no time at all he was being called "Good Roads Campbell." As one result of his efforts Ontario abolished statute labour in 1900. Henceforth it would depend upon publicly-financed professionals to build its roads. The following year the legislature allocated one million dollars to aid the counties in road improvement. The province would pay one-third of the cost if the work was done to government standards. The local bodies accepted the challenge. A total of $3,393,507 (an appreciable sum in that period) was spent on 3,771 miles of roads. Campbell's engineering concepts were beginning to spread.

Ontario had recognized the two basic requirements of the new look in transportation: a professional civil service to administer the highways, supported by plentiful amounts of money. Others followed parallel paths. Quebec anticipated Ontario by two years in creating a formal Department of Highways in 1914. Gradually civil engineers began to supersede amateur builders, and technical reports came to replace proposals which had often been wishful thinking, if not outright political jobbery. The horseless carriage was a far more exacting servant than Old Dobbin. It had great virtues. It was fast and untiring, but to be efficient it had to travel at its natural cruising speed. Twisting, roller-coaster roads suited to a horse's speed bridled the automobile. An animal might welcome a pause as the road turned ninety degrees at the entrance to a bridge; to the motorist it was an inconvenience. He wanted smooth, straight highways with gentle grades and regulated curves.

He also needed a better surface. The speed and weight of automobiles tore up the unimproved dirt roads, and even on "macadamized" highways (as gravel-surfaced roads were popularly and erroneously called) the fast-turning wheels kicked up blinding clouds of dust which coated the traveller and the crops along the roadside and made driving a hazard. On main routes gravel was spread over the dirt roadbed to provide a harder-wearing surface at minimum cost, and oil or tar was used to keep down the dust. Clearly a surface that would stay hard at all seasons and in all weather would be welcome. In 1910 Quebec built a ten-mile concrete road connecting Montreal and the village of Ste Rose. It was the first concrete highway in the province and one of the first in Canada. In the same year construction was begun in Ontario on a concrete highway leading from Toronto to Hamilton. Its completion in 1915 made it the first concrete highway in Ontario and one of the longest intercity concrete stretches in the world.

Such progress was not warmly welcomed in all quarters. Considerable

Yonge Street, Toronto, about 1920.

opposition arose from merchants, livery stable owners, hotelkeepers, and municipal councils who feared they would lose from rebuilding, rerouting, or consolidation of the roads. As a result highway development gathered momentum slowly during the first two decades of the present century. As late as 1918, when the main road west from Toronto had already been paved, the highway east to Montreal (the

CROSS - SECTION OF CONCRETE HIGHWAY

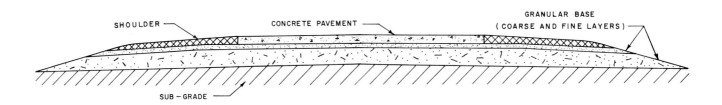

link between Canada's two largest and wealthiest cities) was still macadamized in the nineteenth-century sense—as dusty, muddy or snowbound as when Weller's stagecoaches sprinted or crawled along it.

Then the pace quickened. The number of cars was increasing rapidly, and, thanks to the manufacturers, they were safer, more dependable, and more comfortable. Motors, tires, and brakes were being steadily improved, the self-starter was replacing the crank, and the enclosed automobile, with windshield and side windows of glass, now protected the traveller from dust and weather. People were beginning to take long motor trips for sheer enjoyment. Visitors from the United States were arriving in their cars in growing numbers. Canadians woke up to the fact that the tourist trade could become an important asset, provided the visitors were not frightened away by the roads they met. At the same time commercial highway use began to develop, although the spindly trucks of the time rarely made trips of more than one hundred miles, and buses were used almost entirely for local routes. Between 1920 and 1930 the registration of motor vehicles increased from 408,790 to 1,232,489. Expenditure on roads by all governments also increased three-fold, to $93 million annually. This proportionate rate of growth was never equalled.

Just as important was the change in highway organization. At the close of the First World War, roads were still the responsibility of local municipalities in most provinces; and many of the most important highways reflected the vagaries of parochial planning, shifting direction at every county boundary in order to pass through the county towns. As the cost and importance of highway building grew, however, the provinces assumed an increasingly active role. Their first move was to offer financial assistance to local governments, subject to some control over standards of design and construction. This was not enough. The cost of first-class highways was far beyond the county and township purse. The subsidies grew higher and higher. To pay them the provinces passed legislation increasing the amount of money they could borrow for road construction. They found other sources of revenue, too. Nominal license fees had been collected at first merely to cover the administrative cost of keeping track of vehicle ownership within a province, and leather license plates had been issued to all registered vehicles. As soon as it became clear that the license fees would not only cover costs but provide additional revenue as well, they were duly raised.

Alberta was the first province to levy a gasoline tax: in 1922 it added a two-cent special sales tax to the price of gasoline to provide funds for roadwork. Other provinces soon adopted this method. With the spectacular increase in the number and use of automobiles in the twenties, provincial revenue from motor vehicles (both registration and gasoline taxes) soared almost seven hundred per cent, from slightly more than $6 million in 1920 to almost $43 million in 1930. During the Depression, revenue continued to grow as tax rates rose sharply. Nevertheless, the provinces always spent more on highways than they

Well into the automotive era many city streets remained unpaved. This is Burlington Street, Hamilton, Ontario, about 1925.

obtained directly from those who used the roads. The balance was made up from general funds in recognition that good roads are an economic benefit to all. Eventually each province created a separate department of government specifically entrusted with the task of building and maintaining the growing network of highways.

Some financial support also came from the federal government, which obtains considerable revenue from motorists through excise and sales taxes, and from customs duties on imported vehicles and parts. Historically Ottawa's motor revenue has always exceeded its expenditures: under the British North America Act, roads are primarily a local and provincial responsibility. Nevertheless, the federal government played an important role in encouraging highway development. As early as 1911 Sir Robert Borden had promised federal assistance for roads

Bad as urban roads might be, the countryside could usually provide the motorist with something worse.

The Ottawa-Prescott highway in course of construction, 1920. Paved motor roads spread slowly along the most heavily travelled routes, but during the inter-war years they never amounted to more than a small fraction of Canada's total mileage.

and had introduced a bill to this effect in Parliament, but it failed to pass the Senate. In 1919, however, the proponents of good roads were more successful. The Canada Highways Act set aside $20 million towards the cost of improving and building roads over a five-year period. Federal grants, divided among the provinces roughly on a per capita basis, would meet up to 40 per cent of costs. When it became obvious that the work could not be completed within the original time limit the assistance was extended to 1928.

It is easy to exaggerate the decline of the horse and the spread of good roads in the years between the two wars. Right into the 1940s horses were used in Canada's largest cities to pull milk and bread delivery wagons; on the rural roads of the twenties and thirties they commonly shared the right of way with gas-driven buggies and trucks. And while Canada led the world in some technical phases of road-building, the general level was not always one to be proud of. The state of prairie roads at this time is suggested by a favourite sport: shooting at gophers from the front seat of a moving Model T, as the creatures popped up from their holes in the roadbed. Many a Saskatchewan farmer in the twenties preferred the old grassy trails he had known in the past, for the newly opened roads turned into sloughs after a rain, and in winter the driver was apt to slip off the crown into the ditch. Yet undeniably improvements were spreading. Across the country, highways gradually were straightened and widened. Slopes were cut to a more gentle gradient, valleys were bridged, drainage was improved.

Construction and maintenance equipment grew more imposing and efficient. Early in the century one of the most common implements on western roads was the split-log drag, a primitive horse-drawn wooden frame which evened out the bumps and smoothed the scattered gravel. One advocate of this inexpensive apparatus called it "the first aid to the injured, . . . an implement that will work wonders in a very short time, if it is used at the right time: . . . the 'stitch-in-time' of provincial roads."

But despite the concern of shortsighted economy-seekers, the split-log drag was replaced with more advanced equipment—first with drags made of heavy planks, then with metal devices drawn by tractors. In the early days of wooden drags, competitions were held to encourage their proper use in rural areas. These were almost as popular as the better-known plowing matches, but as the calibre of equipment and professional staff improved, such incentives ceased to be necessary.

Animal power continued in use for some time. In Manitoba a popular implement was the "fresno," a horse-drawn scraper used to keep roads wide and flat. In Saskatchewan during the early twenties, road-builders used an elevator grader pulled by ten horses. The grader scooped up the dirt onto a conveyor belt and dumped it into a waiting wagon—also horse-drawn. Steam was an important source of power for mechanical shovels and rollers in the twenties and thirties, and also for crushing the large quantities of stone needed for better surfacing; but steam engines were too cumbersome for hauling, and only occasionally were small-gauge locomotives used on road projects. The big change in powered equipment came with the advent of gasoline-driven trucks and crawler-tractors. The tractor in particular, powerful and stable under the toughest conditions, transformed the work of road-building in its many guises as grader, earth-mover, and bulldozer. Concurrently mass production methods of paving were developed. Streams of trucks arriving at the construction site loaded with raw materials made the laying of concrete or asphalt surfaces an almost continuous process.

While technical innovations continued during the thirties, the economic depression brought road-building in many parts of Canada almost to a halt. In some provinces even essential maintenance had to be neglected; in others improvements continued, but at a much slower rate. Where funds for hard surfacing were still available, a common method of stretching them was to pave only one side of the highway, as in the days of plank roads, and let travellers in either direction use the better half until they met: then each would keep to the right-hand side. Later, when more money could be found, the other half would be covered over. In parts of western Ontario road-builders experimented with an economical alternative to asphalt and concrete. Their "salt and batter" mixture was a combination of clay, salt, and sand; it produced an almost white road which stayed hard as long as the sun shone, but the salt absorbed moisture and after rain or frost the surface was spotted with pot-holes.

In 1934 Canada had about 409,300 miles of highway and rural road. Much of this network was still unimproved, impassable for anywhere up to five months in the year. Less than a quarter of the total mileage was surfaced, and less than two per cent was paved. As might be expected, most of the improvement that had been made centred in Ontario and Quebec, the richest provinces and the ones with the most automotive vehicles. The prairie provinces, on the other hand, were responsible for more than half the total mileage, yet they had been the last to begin road-building and were the hardest hit by the depression.

Two years later the Canadian Good Roads Association termed the highway situation "still not good": an "eighty mile per hour car" with a "twenty mile per hour operator" was still using a "thirty-five mile per hour road." The Association showed by investigation that thirty per cent of road construction in the preceding twenty years was unfit for the traffic and speed of the times, and proposed some ground rules for the future. It recommended that roads follow the shortest route between two points, rather than the winding trails that originated with the pioneers. Initial construction costs would be larger, but savings would be made on surfacing, guard rails, maintenance, and property. It also recommended a maximum curvature of three degrees, a minimum sight distance of eight hundred to one thousand feet, and a maximum gradient of four to five per cent.

When this pessimistic note was struck, it must be noted, Ontario was already building the Queen Elizabeth Way as a successor to the old paved highway from Toronto to Hamilton. The four-lane controlled-access route—Canada's first superhighway—was opened by King George VI and his Queen, after whom it was named, shortly before war broke out in 1939. At the time there was nothing comparable in the world except for the German autobahns. The Pennsylvania Turnpike, the first modern freeway in the United States, was not begun until 1940.

Obviously the pattern of highway development varied greatly across the nation. Canada would achieve a certain homogeneity in road building in the 1950s; but up to the Second World War, although some construction methods were in general use, each province met the challenge of the automotive age in its own way.

The difficulties of road-building in Newfoundland proved to be second only to those in British Columbia. The island's population was poor, closely controlled and often exploited by the fishing industry, and scattered in many tiny villages along the coast; its terrain was rough and hostile, with rock, forest, muskeg, and quicksand. Here the impact of the automobile was slow to be felt.

Newfoundland

The first cars arrived from the mainland early in the century. For many years thereafter ownership was restricted almost entirely to wealthy residents of St. John's, and licensing was left to its municipal council. Throughout Newfoundland the construction and maintenance of roads, bridges, sewers, and drains were in the hands of local road boards, whose members also were responsible for the abatement of nuisances, the lighting of towns and settlements, the construction and maintenance of public wharves and breakwaters (unless harbour boards were specially authorized to perform this work), the keeping of dogs, and in special cases the relief of the poor!

In 1925, when there were 1,054 cars and trucks on the island, Newfoundland took the first concerted step towards better roads. A Highroads Commission was appointed to plan and carry out construction of new routes and improvement of old ones, and machinery was introduced on a limited scale. The Commission's efforts were concentrated on the Avalon Peninsula, where population density was highest, and by 1930 most communities in that area were linked by road. Some development also began on the Bonavista and Burin Peninsulas, in the Grand Falls and Corner Brook districts, and in the Codroy Valley, although these systems were unconnected. Then the Depression struck. New construction came almost to a standstill. When relief payments were as low as four cents per day per adult, even road maintenance had to be kept to the barest minimum.

In 1934 a Commission of Government took over the island's administration, with a reconstruction scheme which included arterial road construction. The keynote was self-help: the government would provide building materials if the people themselves would do the work. Several communities as a result were connected by road. In 1935 the Commission also embarked on a highway from Port aux Basques to Grand Falls, the western half of a proposed trans-insular route. Some fifty miles were built before war broke out in 1939; then the project was temporarily abandoned. At that point most parts of the island were still inaccessible to automobiles.

Prince Edward Island

The introduction of motor cars to Prince Edward Island was a long and painful struggle. The first machines appeared there in 1907 and 1908. In 1909 the Premier, who drove a horse and carriage and was deathly afraid of automobiles, introduced a bill in the legislature entirely outlawing their use on the province's streets and roads. It passed with only four dissenting votes. The total ban remained in force until a new government under the Hon. A. E. Arsenault took office. By 1914 certain roads were at last opened to motor vehicles for three days of the week, and in 1919 the horseless carriage was finally granted the freedom of the island. When that decision was made the Premier was approached by a good friend and supporter who told him, "You have ruined the Conservative party." Mr. Arsenault replied: "Mr. H——, you are now selling buggies and carriages, but soon you are going to sell motor cars and make more money than ever." His predictions, the former premier recalled later, were both fulfilled.

Prince Edward Islanders had shown greater foresight a generation earlier, when in 1877 they created a Department of Public Works, one of whose functions was to improve road-building techniques. In the same year, long before any other province, they abolished statute labour. Starved for revenue, however, the government was forced to rescind that decision, and statute labour was reinstated in 1879. It was abolished a second time in 1901 in line with the national trend, but again was reinstated in 1912, and continued more or less in force until 1948.

The chronic shortage of funds was relieved somewhat after 1919 when the federal government began making grants. Highway building was stimulated for a time, although the province still had difficulty in raising its sixty per cent of the cost. By 1923 an 850-mile road system —distinctively pinkish in colour from the local sandstone—circled the island and connected almost every town and village. Experiments in hard surfacing began in 1930. Progress was not rapid, but by 1935 the paving extended for fifty-seven miles.

Nova Scotia

The road system of Nova Scotia was well established at the arrival of the automobile. As early as 1850 stage coaches were already serving every important community at least twice a week, and by 1917, when the first motor trucks appeared, there were nearly fifteen thousand miles of roads in the province. On Cape Breton Island, however, routes were so narrow that the trucks could scarcely make their way along them. Elsewhere in the province, as in the Maritimes generally, the slim motor tires damaged the bridges and cut the earth roads so that they turned into quagmires after a rain. Road plows and graders were coming into use, but the small appropriations made for roads precluded macadamizing, or even gravelling. (While it cost about $1,500 per mile to gravel a road, grants for the purpose were often no more than $500 per mile.) In an attempt to preserve what already existed, the Highways Act was amended to prohibit motor traffic between March 20 and May 1, when the roads were particularly soft. Some experts argued for a similar prohibition in the autumn, and for the outlawing of narrow tires on loaded vehicles.

In 1918 the provincial Highways Commission took over from the municipalities the responsibility for construction and maintenance of all highways outside the boundaries of towns and cities. A Department of Highways, with similar duties, was established in 1926. There was a great deal for it to do. More than half the roads in the province were still unimproved; a quarter were graded but unsurfaced; of the remainder, most were simply gravelled. Better, straighter roads gradually were built—though often not without local grief. The little village of West Dublin, south of Lunenberg, rose up in arms when the gravel road reached its outskirts in 1928, for the surveyor had aimed the route clear through the district's most famous rose garden. The neighbours protested, but the foreman uprooted Maggie Romkey's rose bushes. In the next provincial election her son George was elected to the legislature and as his first official act had the foreman dismissed. The road remained. As Highway 31, it still runs in front of Romkey's white clapboard store.

Improvement in Nova Scotia was steady in the thirties, but slow. It was 1937 before the Department of Highways was able to assume even part of the responsibility for snow removal on trunk roads. Until then the first snowfall had ended all departmental labours for the year. Farmers had been left to dig themselves out—or remain snowed in.

New Brunswick

Early motorists in New Brunswick, as elsewhere, had their troubles. The eighty-two-mile journey from St. Stephen to Saint John took ten to twelve hours in good weather. Rain added a hazard that only the most intrepid driver was prepared to risk. Provincial expenditures on roads were extremely modest and confined chiefly to maintaining the existing system. In 1917 public demand for better conditions resulted in the establishment of a permanent roads division under the Minister of Public Works, J. P. Veniot, a well-known figure in the early good roads movement.

In 1920 New Brunswick had 1,371 miles of so-called highway, plus many miles of "branch" roads. During the next five years, with federal assistance, 339 miles of highway were reconstructed to standard gravel type and 215 miles were improved. The highways division was reorganized in 1926. The province was divided into seven districts, each under the control of a district engineer, and the pace of development quickened. Expenditures more than tripled between 1925 and 1929. In 1930 legislation was enacted to build hard-surfaced roads, and four years later the province began a continuing program of road modernization.

Quebec

Some improvements in Quebec's travel conditions antedated the motor age, especially under the Mercier administration of 1887–91, when wooden bridges were replaced in many cases by iron ones. Yet the provincial grant for roads in 1900 was only $5,000, and six years later it was still less than $10,000. In 1907 and 1908 the effects of the automobile began to be felt. The province began making limited grants to municipalities for the construction of roads to provincial standards, for gravelling and macadamizing, and for the purchase of road machinery. Immediate results were not startling: the sums were small, and there was not yet great public interest in good roads.

On some rivers the only way to cross was still by ferry.

A plank road persisted alongside the dike near Laprairie, Quebec, in this period. Its construction was most unusual in that the boards all ran lengthwise.

In 1912 a roads division was formed, under the Department of Agriculture, and directed by a deputy minister. The government was authorized to borrow ten million dollars to assist municipalities in the creation of a highway network. This time provincial officials went round the countryside to explain the importance and value of good roads. They were so successful that local councils had to be asked to restrain their new enthusiasm. In the first six months of 1913 road expenditure under the Department of Agriculture amounted to $1,592,393.

The province's first auto route, the Boulevard Edouard VII, was built with a macadamized surface in 1912–13. It ran thirty-four miles from Montreal, via Laprairie and Napierville, to the New York border at Rouses Point. There it joined the historic route to the south along the Richelieu Valley. Macadamizing began about the same time on the trunk route between Montreal and Quebec, and was completed in 1918.

After the First World War Quebec undertook large-scale repairs and construction. Rural roads were improved to speed farm products to urban markets, and arterial highways were built to the borders of the United States and neighbouring provinces. In the first decade after the war the number of miles of good roads quadrupled. In the same period, however, motor traffic multiplied by six, and the number of cars entering the province from outside increased thirty-fold. In 1928 the Legislature authorized expenditure of $17 million in road construction; in the following year the reward was evident when American tourists spent $61 million in the province. The emphasis on roads for tourism continued in the 1930s, despite the Depression. Most notably, a tourist route 561 miles long was built to the Gaspé. Some of its long, steep, gravel-surfaced grades were a severe test of the automobiles of that era, but the scenery was, and is, amply rewarding. Attention was also paid to expanding existing routes to accommodate greater traffic, and to constructing roads which could stand up to heavy transport. From Montreal to Sorel the highway was paved with concrete between 1933 and 1940.

Under the guidance of its Instructor in Roadmaking, and with additional funds allocated under the Highway Improvement Act, Ontario continued to assist its municipal roadbuilders throughout the early 1900s. In 1912 a new element was added with a five-million-dollar grant for northern roads. About the same time a Public Roads and Highways Commission was appointed by the Legislature, and in 1915 the Department of Highways was created under the Commissioner of Public Works. Municipalities continued to be responsible for the roads within their jurisdictions, but the Department of Highways assumed some responsibility for a limited number of trunk routes. A subsidy on their costs, beginning in 1917 at twenty per cent, rose gradually to forty, sixty, and eighty per cent; finally the province assumed all the charges in 1935 and the modern Highways Department was born. Ontario owes much to the vision of two of its early deputy ministers of highways, W. A. McLean (1914–23) and R. M. Smith (1927–43). During the years of their tenure thousands of miles of highway were surfaced; new

Ontario

The motor transport industry was young but growing. Private truckers carried soldiers to camp near Brantford, Ontario, in 1927.

roads were opened in central Ontario; colonization routes were pushed across the rocks and muskeg of the north.

One of the most familiar of Ontario's innovations is the dotted white line which marks the centre of many paved highways. In 1930 a young engineer, J. D. Millar, was in charge of construction along a stretch of Highway No. 2 near the Quebec border. The road was flat and straight and when fog blew in, as it often did from Lake St. Francis, motorists had a difficult time keeping out of the ditch. To guide them, Millar painted white dots every three hundred feet down the centre of the road. White paint till then had been reserved for solid lines at uncommonly dangerous corners. When Millar's superior saw the dots he ordered them blocked out. "What do you think we're going to do?" he asked. "Paint white lines on all the highways of the province?" Yet within three years Millar's dotted lines had become standard throughout the North American continent.

Two further innovations—the twelve-inch traffic light lens and the thirty-foot depressed boulevard between opposing traffic lanes—were part of planning for the Queen Elizabeth Way. This early superhighway also had the distinction for a time (until wartime rationing of electricity) of being the longest continuously lighted rural highway in the world. Its first stretch, from Toronto to Hamilton, was completed before the outbreak of the war; a continuation to Niagara Falls was graded by that time, and was considered so important that the federal government gave priority to finishing the job by the end of 1940. The Queen Elizabeth Way was used so heavily by war industries in St. Catharines, Hamilton, and Toronto that by the time peace returned in 1945 its pavement was almost completely destroyed.

Floods appear to have been a special turn-of-the-century problem in Manitoba. In 1902 alone, 230 bridges were washed out and many roads badly broken. Calls arose for "proper" road construction and, as farm machinery grew heavier, for concrete bridges as well. Extensive use of gravel on the roads began in 1909. The following year the Manitoba Good Roads Association held its first convention, and the province appointed its first highways commissioner.

Manitoba

Municipalities were still responsible for roads within their boundaries, but they received two measures of provincial assistance in 1912. The first was permission to issue debentures for road construction under specific conditions. The second was a government allocation of $200,000 to pay two-thirds of the cost of improving certain roads recommended by the highway commissioner. One of the first stretches to be given a new look was the old Red River Trail from Winnipeg south. It was asphalted in 1913. Provincial participation was further broadened in 1914 by an act authorizing the government to raise capital funds for road construction up to $2.5 million.

Efforts were made during the First World War to show farmers that good roads meant dollars in their pockets. The rate of construction continued to grow, particularly as improved machinery was introduced. By 1920 the Wilson drag with tractor power was replacing the split-log drag in some localities, and in districts where a gravel quarry was convenient one might see a steam shovel in use and a small industrial railway bringing gravel to road-workers. In 1925 there were 1,600 miles of "superior" roads. About this time Manitoba began building modern roads into the northern and central areas of the province to open these mineral-rich districts to exploitation; and while the Depression slowed road-building in general, these resource projects were advanced by relief work.

When Saskatchewan came into existence as a province in 1905 it faced enormous problems of building and maintaining roads between centres of population widely dispersed over a vast area. In its first year it spent only $50,000 on roads and bridges (considerably less than the cost of building one mile of the Trans-Canada Highway half a century later). There were perhaps a score of motor cars in the province at that time. By 1911 there were 1,300, and public demand for improved roads was increasing. In 1912 a Board of Highway Commissioners was established to administer capital expenditure on main roads, bridges, and ferries. The province began to develop its market routes, sharing construction costs dollar for dollar with the municipalities up to a fixed maximum. In the period from 1912 to 1914 the average annual expenditure was two million dollars. Construction slowed down with the outbreak of war, however.

Saskatchewan

The "Bennett buggy"—named after Canada's then Prime Minister— symbolized the Depression in the west, where many a farmer returned to horsepower because he couldn't afford gasoline.

In 1917 the Board was replaced by a Department of Highways under a cabinet minister. Two years later planned development of highways began when an outline of proposed trunk roads was submitted to the federal government for approval under the Canada Highways Act. The $1.8 million forthcoming in federal funds established the Department of Highways, its standards, and its practices.

By 1921 Saskatchewan had 61,184 motor vehicles—or one vehicle for every 12.4 persons, the highest rate in Canada. Public expenditure and effort increased accordingly. During the five years up to 1925, two thousand miles of road were graded, of which forty-nine miles were gravelled. By 1930 there were 4,682 miles newly graded, and about 1,900 miles of this were gravelled. With so extensive a system and a relatively small population, there was no money for asphalt or other hard surfaces.

In the Depression, construction came to a halt. Even maintenance had to be deferred; as a result, in many areas the original investment in road improvement was entirely lost. Automobile registration dropped off as well. Lacking the money to buy gasoline, the farmer often left his car to rust, or else took out the motor and hitched the body behind a plow horse.

Alberta

The roads of Alberta—what there were of them—were transferred in 1905 from the government of the North West Territories to the new province's Department of Public Works. There were few immediate changes in the methods or rate of construction. In 1908 the Department reported great success with corduroy roads, and in 1914 road builders were still quite happy with widths of sixteen feet. Anything more was considered a luxury.

Post-war growth in motor travel led to public demand for properly constructed highways and in 1923 a Highways Branch was created in the Department of Public Works. Construction and maintenance of highways became the direct responsibility of the province, but local and district roads were left to the municipalities. The first Alberta road map appeared in 1924. It described the highways by colours, not numbers; road signs posted on telephone poles along the 250 miles of highway also used colours. Route numbers were not adopted until 1929.

The Calgary-Edmonton Road has long been the most heavily travelled in the province, connecting as it does the two major centres of population. Highway No. 2 was begun in 1926–27, closely following the old trail cut by the Rev. John McDougall in 1873. Five years later the first stretch of asphalt in Alberta was laid over thirty-one miles of the highway between Red Deer and Ponoka. By the end of 1932 the province had a total of 2,960 miles of main and secondary highways, approximately five hundred miles being gravel surfaced and sixty-seven asphalt.

In 1889 British Columbia took its first survey of road mileage. Men on bicycles fitted with cyclometers pedalled wearily through valleys and over mountains. They logged a total of 5,615 miles of wagon roads and 4,400 miles of trails.

British Columbia

In 1902 the province's first gasoline-powered automobile coughed its way down the main street of Victoria as part of the May 24th celebration of the late Queen's birthday. Two years later, when auto licensing was initiated, total revenue from this source amounted to $36. Yet in 1908 British Columbia announced plans for a new approach to highway building, and created a Department of Public Works to undertake it. In 1918 the provincial road administration was further reorganized: there were no longer local road superintendents in almost every electoral district; now supervision was entrusted to eight district engineers, each in charge of a large area. The engineers were trained technicians, and the reports they submitted were based on economic and professional points of view rather than political ones.

By this time road machinery, which in 1916 had consisted of twenty-five graders, fourteen wagons, a small steam roller, a traction engine, a rock crusher, and some other equipment, had grown to a value of about $250,000. With its professional engineers the province began forming a comprehensive plan to connect all major centres of population and industry. It was recognized, however, that a goodly portion of the available money would still have to be spent on wagon roads to the outlying districts, although some of these rough routes had been built through thinly settled regions solely to serve one or two families.

By 1929 there were 18,200 miles of roads in British Columbia, and 2,300 miles of trails. In the thirties some projects were carried out—notably the Big Bend Highway through the Selkirk Range—but in general the pace fell off, and little could be done during the war years which followed. It was 1947 before British Columbia, and many other provinces of Canada, moved back into high gear on the roads.

At left, survey teams map the future route of a motor road through the Rockies from Banff to Jasper. The result a few years later was some of the smoothest driving then available—through some of the ruggedest country.

In the late thirties Canada's first super-highway imposed a new geometry on the landscape.

CONSTRUCTION PROGRESS

Well into the twentieth century, roads were built without benefit of the new mechanization which was transforming their use. Animal power broke the prairie sod in Saskatchewan before the First World War, and levelled the dirt track which resulted. As late as 1932 dust-coated teamsters still hauled fill by horse-drawn wagon to build the approach to Outlook Bridge (below).

Crushing gravel for the roads of Oxford County, Ontario, in 1909.

The simple wooden drag was common in the west in 1915. With it a farmer could be expected to maintain the roads in his district in fairly smooth condition.

This is road-building by statute labour in New Brunswick in 1917—unskilled workers forced to leave their regular occupations for a few days each year in lieu of paying taxes. Progress was strangled by this inheritance from the colonial era.

The same province a few years later—road-building had accelerated under a permanent organization of qualified engineers, regular construction crews, tax-based revenue, and mechanization.

We really did
make some
improvements.

We attacked this culvert   and   reduced it to this.

Then

fter passing thru this stage   the above result was obtained.

A page from the photo album of a young engineer in Saskatchewan in 1923.

Crushing gravel for eastern roads at the end of the First World War. The gasoline engine had begun to cut time and costs of haulage.

Gradually, road construction
equipment grew bigger, more
complex, more powerful. An early
example of mechanization was the
use of a steam traction engine to pull
a train of wagons in Ontario in 1912.
By this means, it was reported, "the
cost of hauling gravel and stone,
especially for distances over one mile,
can be greatly reduced." The
caterpillar tractor, shown at left
coupled to a grader, and the steam
roller (lower left) became common
sights along Canada's improving
highway network.

Street paving in the days of the five-cent cigar was a far cry from the streamlined operation of today. This early steam-powered mixer speeded up the first stage of the process, but the concrete still had to be poured (below) one wheelbarrowful at a time.

Above right: Heavier, more efficient drags and tractor power were transforming highway maintenance in the 1920s.

Near Orillia, Ontario, in 1935, Highway 12 was paved by a distributor, harrow, and grader operating as a unit. This section was done one side at a time.

13

Opening the Last Frontiers

WITH THE OUTBREAK of the Second World War, road-making in Canada was set back for six long years while men and materials were diverted to more pressing needs. There was one epic exception. When Canadians spoke of "The Road" in 1942 and 1943 they meant a highway which few of them would ever see: a twisting, gravelled route built in a matter of months over five mountain ranges, 129 rivers, and thousands of streams—the 1,523 miles of the Alaska Highway.

Nearly half a century had passed since the first mass invasion of Canada's far northwestern corner. Back in 1898 there had been no roads of any description. No matter—in the rush that year to the gold fields of the Klondike, men were ready to endure any hardship. Most set out from Skagway, on the Pacific, along a treacherous mountain trail beset with streams, boulders, ice, and slides, and lined with the bodies of dead horses and abandoned equipment. On the Chilkoot Pass they toiled in an unbroken black line against the snow, man after man carrying supplies on his back up a slope too steep for pack animals. The descent that followed was even more difficult. Then came the hazardous sub-Arctic rivers, turbulent and rocky, bearing ice floes which could easily crush the crudely-built boats and rafts. Many a man turned back. Hundreds, probably thousands, of ninety-eighters died of exposure and starvation. And when the bubble burst, and the surface gold had been scraped away, most of the miners departed in a stampede that was worse than the rush in.

Some remained, however, and for them the federal government began building roads. By 1913 the Yukon Territory boasted some five hundred miles of good wagon routes, radiating from Dawson and Whitehorse and covering the 330 miles between. Still there was no overland route to the rest of Canada. Access to the Territory was by boat down the Yukon River, or by railway from the American port of Skagway over the White Pass.

Seven hundred miles to the southeast, in Alberta, a primitive road network was slowly expanding into the Peace River country. The Edson Trail was widely advertised as the gateway to that rich valley. In theory it was served by stagecoach; in fact passengers who paid the $60 fare reached their destination, if at all, only after walking, wading, and swimming most of the 230 miles.

In the United States agitation had begun for a road to Alaska. Its proponents even had a slogan: "Seven million dollars purchased Alaska for the States, and seven million more will make Alaska one of the United States." Some experts envisaged a road up the Pacific coast from Washington through British Columbia; others preferred the "back door" approach through northern Alberta and the Yukon. The most enthusiastic saw the projected highway as the northern segment of a single continuous road from Tierra del Fuego to Alaska. But during the twenties there were other demands, and the Depression of the thirties damped all planning.

So matters stood on December 7, 1941, the day the Japanese bombed Pearl Harbor. Overnight, Alaska assumed a new importance. There seemed a possibility—remote perhaps in retrospect, but a frightening possibility then, when the forces of the Rising Sun were sweeping over

The Alaska Highway has opened the far northwest to the tourist and industry. This stretch lies at Mile 707, Yukon Territory.

There were no roads in 1898 when the gold seekers outfitted at Edmonton on their way to the Klondike. Both dogs and horses were used to transport supplies, including a year's supply of food.

Spring thaws played havoc with the overland trail from Dawson to White-horse, the artery of Yukon traffic half a century ago.

Right: The Royal Mail Stage leaves Whitehorse for Dawson on May 10, 1916. It carried passengers as well as mail and was in fact the only public transportation to Dawson. In winter, sleighs were used (below). Travel in the Yukon retained a nineteenth-century flavour well into the present century for want of adequate roads. Taverns along the way featured moose and caribou steaks; meals were $1.50 each, beds $1.00 a night.

the Pacific—that the Japanese would thrust up the Aleutian Islands into Alaska and so invade the North American continent. The northern territory was designated a front line of defence. To supply the troops marshalled there, and a chain of northwestern airports, the Canadian and American governments decided to build a highway over the rough "back door" from Dawson Creek, British Columbia, to Fairbanks, Alaska. Work began the next March, before the frost had left the ground.

When war broke out, Dawson Creek had a population of eight hundred. The impact of The Road was like a second Klondike gold rush. Within weeks twelve thousand American soldiers arrived by railway. Quonset huts and supply dumps sprang up around the town. Water lines and a sewage system were built. Prices skyrocketed. Soon the troops were joined by thousands of civilian workers. "This is no picnic," said the poster advertising for workmen, but the pay was good and there were plenty of willing hands. Inevitably, bootleggers and gamblers followed. The police had a busy time, made no easier when the local jail burned down.

The road-builders launched their assault on the northern wilderness at several points. Quickly they spread out from Dawson Creek to Fort St. John and Fort Nelson. At the other end they worked from Fairbanks, Watson Lake, and Whitehorse. Docks at Skagway and Anchorage were piled high with gasoline drums, trucks, bulldozers, cement mixers, and portable generators. Equipment for a single regiment in this operation included 20 heavy and 24 light tractors and bulldozers, 6 pulled road-graders, 3 patrol graders, 6 rooter ploughs, 6 twelve-yard carrying scrapers, 93 half-ton dump trucks, 1 six-ton prime mover truck, 7 four-ton cargo trucks, 9 two-and-a-half-ton trucks, 25 jeeps, 10 command cars, 1 sedan, 12 three-quarter-yard pick-up trucks, a truck train, 2 half-yard shovels, pontoons, concrete-mixers, compressors, ploughs, gasoline-driven saws, pile drivers, portable electric generating units, and electric welders.

No time was wasted on niceties of engineering. The road was looped over hills to save blasting benches, and rammed through the forest. A twenty-ton bulldozer led the way, with others following on either side to clear debris. Men strung out thirty to fifty miles in the rear wielded saws, picks, and axes as they chopped down trees, built culverts, cleared slash, and hauled gravel, while waterboys quenched the workmen's thirst from galvanized pails. Overhead, aerial reconnaissance planes took photographs of the next stretch of muskeg and forest to be attacked. The muskeg—much of it seemingly bottomless—was largely avoided by changes in direction, so effectively that although muskeg predominated in one 265-mile stretch, only four miles of it were actually crossed. There, rock fill was dumped into the spongy muck and topped by many feet of brush, corduroy, and gravel.

Temporary bridges were built from local timber or on pontoons, sometimes over streams two thousand feet wide which would become raging torrents in the spring. Permanently frozen ground was left undisturbed. Work continued seven days a week, day and night, in temperatures which ranged from sub-tropical heat to fifty degrees

JUNE 15 42

THIS IS NO PICNIC

WORKING AND LIVING CONDITIONS ON THIS JOB ARE AS DIFFICULT AS THOSE ENCOUNTERED ON ANY CONSTRUCTION JOB EVER DONE IN THE UNITED STATES (FOREIGN TERRITORY. MEN HIRED FOR THIS JOB WILL BE REQUIRED TO WORK AND LIVE UNDER THE MOST EXTREME CONDITIONS IMAGINABLE. TEMPERATURE WILL RANGE FROM 90° ABOVE ZERO TO 70° BELOW ZERO. MEN WILL HAVE TO FIGHT SWAMP RIVERS, ICE AND COLD. MOSQUITOS, FLIE AND GNATS WILL NOT ONLY BE ANNOYIN BUT WILL CAUSE BODILY HARM. IF YOU ARE NOT PREPARED TO WORK UNDER THESE AND SIMILAR CONDITION DO NOT APPLY
Bechtel-Price-Callahan

Until The Road was built, this was the fastest means of overland travel in the north.

below zero. At times men worked sixty hours at a stretch. But wise-cracks and humour pervaded the route. Like their brothers-in-arms in the trenches, these soldiers tacked up local signs recalling familiar scenes back home: a pathway through the woods might be named 42nd Street or Broadway; arrows pointed to imaginary nightclubs, or to fictitious stores stocked with unavailable produce.

On November 20, 1942, The Road was ceremonially opened at Soldiers' Summit, overlooking Kluane Lake. Canadian and American officials joined in the ceremony, and the two national flags were hoisted as the band played "God Save the King" and "The Star-Spangled Banner." When the red, white, and blue ribbon had been cut, the first through convoy of trucks from Dawson Creek pushed on to Fairbanks. Ten thousand soldiers and six thousand civilian workers had united to produce a twenty-four-foot roadway at an average speed of eight miles per day, completing in eight months a stupendous piece of construction that would normally have taken more than five years.

The Alcan Military Highway, as The Road was known officially during the war, was still far from perfect. During 1943 and 1944 work continued, installing permanent bridges and culverts, lowering grades, widening the bed, laying gravel, to make a highway usable

all year round in any weather. Six months after the end of the war the United States, which had borne most of the $130 million construction cost, turned over to Canada the 1,221 miles from Dawson Creek to the Alaska border. Then began a third stage of development. Canadian Army Engineers regraded and widened The Road during the next eighteen years until it ranked as one of the finest gravel highways in the world. By 1964 it had become the economic lifeline for development of a vast, rich area. In comparison its military importance had waned, with the result that maintenance was handed over by the Army to the civilian engineers of the federal Department of Public Works. One of their first projects was to study rebuilding the highway, possibly with paving. British Columbia was prepared to assume responsibility for the portion within its own boundaries—once the gravel road was paved to provincial standards. The Department of National Defence had in fact paved the first eighty-three miles, from Dawson Creek to Fort St. John, and had turned that stretch over to British Columbia;

For 1,221 miles the slender line of the Alaska Highway links the scattered communities of Canada's far northwestern frontier.

About 100 miles west of Whitehorse, Yukon Territory, the Haines cutoff leaves the Alaska Highway on a mountainous journey south to the Pacific.

surfacing the remaining distance to the province's high standards, however, was to be a gigantic, costly task.

Maintaining the Alaska Highway has proved to be an engineering feat almost as strenuous as the original construction. Nature's counterattacks never stop. They are particularly fierce in June, when rivers and streams, fed by mountain glaciers, swell in flash floods that sweep trees and boulders ahead of them. In a matter of minutes whole sections of roads can be washed out. In the spring, too, masses of ice weighing up to one hundred tons surge down the rivers, slamming against bridge piers with terrific force. Then there are landslides. Without warning, hillsides slip, blocking the highway or catching a stretch of it up and carrying it away. Slides also attack bridges: the 2,130-foot suspension bridge over the Peace River was the pride of the highway until a land shift in 1957 moved the cable anchor and the span collapsed. It was replaced with a cantilever structure.

During the winter, snow and blizzards demand clearing by a fleet of heavy ploughs. The cold can be extreme: temperatures of thirty-five to fifty-five degrees below zero are not uncommon and can continue for days. Yet summer is so short that construction and maintenance must go on during the winter months. When the Donjek River bridge was built near the Alaska border the temperature fell more than fifty degrees below zero, causing bulldozer blades to crack like glass and steel girders to crystallize and break.

Thousands of tourists who have travelled The Road in summer or early fall remember a much less romantic menace. The highway was built on a gravel base covered with clay. In dry months truck wheels pounded the clay into dust which rose in thick clouds, coating travellers, obscuring vision, and causing accidents. The gravel base, thus exposed, proved a rough road in need of constant grading, but the bumps were no worse than the dirt. Yukon businessmen took to canning "genuine Alaska Highway dust" for southerners to take home as souvenirs.

The remarkable fact is that the Alaska Highway could be a tourist route at all. Within a few years of the war it was being travelled by vehicles from every province and state. Yet it was not easy to reach— Dawson Creek is 475 miles by road from Edmonton—nor was it easy to travel. The approach roads were gravel, and often difficult to negotiate in wet weather. The authorities were quick to warn of difficulties: accommodation and service stations were limited, telephones and garages far apart. Only first-class cars should attempt the journey, an official publication warned, "for this is virgin territory, the new land of North America, one of the continent's last frontiers. . . . The trip is no push-over. It is not for 'sissies.' " Despite this pessimism motorcyclists reached Fairbanks, and so did the odd Model T Ford. A regular modern bus service was initiated with overnight stops at lodges. And in this wild country the Royal Mail was brought to some five thousand people along The Road by truckers who hewed to so tight a schedule that deliveries were rarely more than five minutes late.

The Alaska Highway proved spectacularly that roads could provide an economical and successful means of penetrating the north. It has opened up previously inaccessible areas for mineral exploration, big-game hunting, and forest industries. From it extend hundreds of miles of side-roads—some little better than trails, others gravelled and improved to carry traffic to mining camps. The most famous of these at first was the Canol Road, which paralleled the five-hundred-mile pipeline from the oil fields at Norman Wells, high in the North West Territories. At the end of the war the refinery was moved from Whitehorse to Edmonton, the pipeline dismantled, and the Canol Road abandoned. Other routes appeared, however. From Watson Lake, a branch road was built to Tungsten, where a large high-grade deposit of tungsten-bearing ore had been discovered; the mineral, rare in the western world, is used to harden steel for missiles and other purposes. A few miles beyond Watson Lake another road branched south to the rich asbestos mines of Cassiar, B.C. At Whitehorse a new road was pushed north to Dawson City; the old gold-rush town is now served by fleets of transport trucks. Still further up The Road, at Haines Junction, begins a spectacular passage south through British Columbia and the Alaska Panhandle to the seaport of Haines. Much of the route was pioneered in '98 by Jack Dalton, who herded cattle along it to feed the hungry miners in the Klondike. His old trail was followed in 1942 to rush supplies for building the Alaska Highway, and though it climbs well above the tree-line in the mountains, it proved too valuable to be abandoned afterwards. Besides these well-established routes, many a trail has been opened to good fishing and hunting territory. From the highway, too, many oil prospectors have set out on hunts of a different sort, exploring as far as the Arctic coast.

The vital importance of building other northern roads was outlined dramatically in 1956, in briefs presented to the Royal Commission on Canada's Economic Prospects (the Gordon Commission) by the Commissioners of the Yukon and North West Territories. These two territories comprised one-third of Canada's area, an untold portion of its mineral wealth—and only about twenty-five thousand people, among

Grading a tote road near Kemano, B.C. The town was only a pinpoint in the wilderness some 400 miles northwest of Vancouver until 1951, when it became important as the power site for a giant aluminum smelter.

them Indians, Eskimos, fur-traders, miners, missionaries, fishermen, scientists, and government officials. No one, the Gordon Commission was told, had any idea how much wealth was locked in this northland, but obviously it was going to grow increasingly important as resources in other parts of the world were depleted. At the same time, nuclear energy promised new comfort for life in the Arctic. Yet progress was impossible without adequate transportation, and only a minute fraction of the vast area was so served. As a result it was difficult and costly to live in the north, and impossible to do more than peck at its mineral-rich fringes.

Obviously airplanes, boats, and railways would play important roles in the development of the north. But roads would have a special function. Roads were relatively inexpensive and easy to build. Their plan could be flexible, altering to meet changing requirements. They could carry large volumes of heavy traffic all year round. Once the non-renewable resources of an area were finally exhausted the roads could

At Fort Smith, capital of the North West Territories, the roads would amaze travellers of an earlier time. The horse and the auto co-exist more amicably than they ever did farther south; here both are necessary.

Warning signs anticipate the laying of tracks for the Great Slave Lake Railway through the Hay River district, North West Territories. Meanwhile local transportation is by the gravel road, which must be kept graded and in good repair.

be abandoned without loss, having paid for themselves many times over. No one imagined a network of multi-lane highways linking the far northern communities. What was needed was a number of relatively low-standard resource and access roads built to accommodate the most rugged kind of traffic. In this the federal government could expect to be joined by companies building private roads to exploit a specific mine or other resource.

At that time the most important road in the North West Territories was the Mackenzie Highway, a joint project of the federal and Alberta governments which ran from Grimshaw, in the Peace River Valley, north to Hay River on the shores of Great Slave Lake. The volume of traffic passing over the 384-mile route had been rising steadily for some time. Along its length moved supplies, machinery, and equipment for the growing mining communities. Southward flowed the products of the mine, fur pelts, and fish from Great Slave Lake. Clearly the highway would soon be unable to meet the growing demands upon it.

In 1958 Ottawa announced plans for a $31 million expansion of roads in the Yukon and North West Territories. In the Yukon this included a development road from Flat Creek, 25 miles southeast of Dawson, to the Eagle Plain Oil Reservation, and from there another 160 miles north to Fort MacPherson, which stands at deep water near the mouth of the Mackenzie River. In the North West Territories the major projects included improvement of the Mackenzie Highway and its extension around the west end of Great Slave Lake to Yellowknife on the north shore. From there the road would be carried north to Great Bear Lake and east to Fort Reliance. Smaller roads would be built to facilitate transportation up the Slave River, and to encourage logging elsewhere. By the end of 1963 the two territories had nearly 1,700 miles of development road in operation, not including the Alaska Highway, and more were under construction.

As administrator of the Yukon and North West Territories, Ottawa clearly had a responsibility for building roads within their boundaries. South of the sixtieth parallel Canada's development was also blocked by want of roads, but the provinces, which were here responsible, lacked funds to extend the highways into thinly settled areas. To help them the federal government proposed in February 1958 a co-operative arrangement for building "Roads to Resources": it was to consist of a fifty-fifty sharing of costs between provincial and federal governments to build development roads which would assist in the exploitation of minerals, timber, fisheries, and tourism. These were not to be super-highways but pioneer routes, opening up areas which would otherwise remain closed for years to come. The initiative was to come from the provinces, which would have to formulate plans and be responsible for eventual maintenance. The response was immediate. Final agreements between Ottawa and the provincial capitals covered a joint investment of $145 million in 4,100 miles of roads—a total mileage which, if put together in one continuous strip, would run almost from St. John's to Victoria.

As work began on the new roads to resources, it was anticipated the projects would require anywhere from five to eleven years to complete. Many of them would be built through country where muskeg and extreme cold presented special problems which Canadians have had to learn to overcome. Yet it was clear that this was only the beginning of a long-term, large-scale assault on Canada's last frontiers.

A convoy of heavy tractors hauls supplies towards the iron ore deposits of the Quebec-Labrador border area. The road they follow now carries the tracks of the Quebec North Shore & Labrador Railway, which links the mines with the St. Lawrence at Sept Iles.

14

Canada Drives Ahead

DURING THE WAR YEARS—when little more could be done than to extend a few key highways such as the Queen Elizabeth Way, build supply routes to military bases and airfields, and try to maintain the routes which already existed—Canada's highway builders were busy planning for peace. With the resources available they studied future needs and laid out possible new developments. It was well that they did so, for when peace came they were faced with an imposing task. For nearly two decades, during depression and war, Canada's road system had been starved. A considerable backlog of essential repair, improvement, and extension had accumulated. The equipment to carry out the work was outdated and in need of replacement. There was not enough money immediately to meet all the needs of a nation retooling after battle. Yet at the same time the population was swelling with immigration and the "baby boom." Gasoline was off ration, and the manufacturers were turning out new cars that were bigger, faster, more powerful, and better suited to long-distance travel than any road vehicles in history.

The road-builders made valiant efforts. In the five years up to the end of 1951 Canada spent a record $1.38 billion on maintaining, extending and improving her highway network. Of this amount well over half was devoted to new construction and major improvements. The total mileage of surfaced highways rose from 145,809 to 173,232. About three miles out of every ten now fell into this category. Most were surfaced with gravel or crushed stone, but the amount of paving also grew by more than 7,100 miles. These figures, moreover, do not reflect improvements in quality within a given class. In general, roads were built better and wider, more suited to modern traffic conditions, better designed to withstand the heavier loads and increased traffic volume.

On a wide front Canada had moved into second gear on the way out of the horse-and-buggy age. But the automobile manufacturers were in high gear. New cars and trucks poured out of Windsor, Oshawa, and Hamilton at a rate the road builders could not hope to match. The number of motor vehicles per mile of surfaced highway rose sharply from 12.6 in 1947 to 16.6 in 1951. "In other words," the Dominion Bureau of Statistics reported, "an 18.8 per cent increase in the mileage of surfaced roads during the five-year period has been virtually swamped by a 56.5 per cent increase in the number of motor vehicles registered." By 1956 the number of motor vehicles per mile of surfaced highway and rural road had reached 20.2—almost twice what it had been at the end of the war. In the next year efforts of the road-builders began to be felt. That year the figure dropped to 19.3, representing a small thinning of traffic.

By that time, however, there was an even more serious problem to be met. Congestion on the highways was dwarfed by that in the cities. The streets, particularly in downtown areas, had been laid out for horse-drawn carriages and carts at a time when most people walked the relatively short distances to work or to the corner store. With the post-war growth of large suburbs and "dormitory areas," workers were driving up to twenty-five miles each morning from home to office or factory; shoppers drove downtown or to the plazas, which were supplied

with merchandise by truck. More than half the movement of automotive traffic was taking place within built-up areas. With so many thousands of cars using the main streets, traffic jams became part of the way of life: people *expected* to have to crawl bumper-to-bumper for miles in the morning and evening rush hours. There was considerable fear that in the end the modern city would simply strangle.

The busiest stretch of road in Canada was in the north-central portion of Metropolitan Toronto, along Highway 401. This four-lane divided controlled-access highway had been designed after the war to carry traffic around Toronto as part of a trans-provincial rural route. No one then had any idea how fast the city would grow. By the time the Toronto "by-pass" was completed in 1956, the suburbs had leap-frogged over it. Highway 401 immediately became a major artery for traffic between points within the metropolitan area. One out of every ten cars in the district used the road twice a day, on the average for a trip of about four and a quarter miles, and during the peak hour about ten thousand cars and trucks entered its lanes. By 1959 that figure had more than doubled to 22,500 vehicles—and plans began for the great expansion of a stretch of highway completed only three years earlier. The result was the most advanced highway design in the country, a complicated series of interchanges feeding into twelve lanes along a seventeen-mile stretch of freeway. The planners called for six central lanes, three in each direction for traffic travelling relatively long distances, and six outside lanes, three in each direction for commuter and short distance traffic. One piece, near Yonge Street, would have

To handle the concentration of urban traffic, today's most advanced highways are being built through the centre of the city. When the aerial photograph at the top of the opposite page was taken in 1959, Highway 401 in Metropolitan Toronto was only three years old. It was already congested with cars and trucks. Over the next seven years planning and construction converted the same stretch to twelve traffic lanes (below), adequate to carry the flow smoothly. A short distance to the west Ontario highway engineers designed Canada's most complex interchange (right) where the highway meets the Spadina Expressway. It occupies 150 acres.

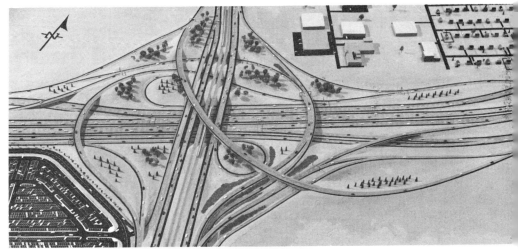

14 lanes. Throughout construction a minimum service of four lanes was to be maintained. The first completed section was opened in 1965. The cost *per mile* of this project—admittedly built through expensive land and to a complex design—was estimated at upwards of five million dollars, or about two and a half times what it cost to build the entire Cariboo Road through the mountains a century earlier.

While larger than any other, the Toronto freeway development was typical of a growing preoccupation among road-builders. During the 1960s engineers began planning and building multi-lane expressways through every major Canadian city. Traditionally, the highway had existed to speed the traveller through the countryside on his way between major centres of population; now it was being extended into the heart of the city to carry traffic at sixty miles an hour between the suburbs and the centre. These projects were invariably expensive. They passed through areas already built up with industry, business, or houses, rather than through open countryside; they had to be built to maximum standards to meet the requirements for safety, speed, and traffic volume. By the mid-sixties nearly forty per cent of the total funds spent on roads in Canada each year was being devoted to projects in municipal areas. The cities, which depended on property taxes for revenue and had no means of taxing road users, could not meet the growing costs alone. Increasingly, provincial governments developed policies of assistance for urban expressways, paying up to seventy-five per cent of the cost of building and maintaining those which connected with provincial highways.

Outside the cities the highways also made bold advances. The Canso Causeway, deepest ever built by man, joined Cape Breton Island to the mainland in 1956. Nine years later work began on a far more ambitious causeway to Prince Edward Island, a land link dreamed of, but seemingly unattainable, for decades. At the other end of the country British Columbia opened a new road into its rugged central interior from Bella Coola, on the Pacific, between the formidable peaks of the Coast Range; it followed a route known for generations but until 1955 impassable to anything but pack trains. Also in the sixties British Columbia completed a stretch of trans-provincial highway between Salmo and Creston which rose more than a mile above sea level: the highest arterial route in Canada.

Further assaults were made on the far north. By 1965 a telephone line had been built to the mouth of the Mackenzie River, and a road planned along the east bank which would replace the region's scattered winter roads and extend the Mackenzie Highway to Franklin's Polar Sea. In the east, Labrador was opened following the discovery of iron ore there. Several towns sprang up in the interior, among them Labrador City, headquarters of the Iron Ore Company of Canada, which could boast paved streets in the midst of the wilderness. Before long Labrador had developed a system of gravel roads linking mining, power, and transport centres. The fishing stations and ports on its long coastline were still unconnected as this was written, but local trails led from many of them, and the roads were evolving, repeating the history of other Canadian regions.

During the post-war years southern Ontario was fitted with a new "main street" running from Windsor eastward to the Quebec border: the 510-mile Macdonald-Cartier Freeway (Highway 401), like other freeways so named because of its freedom from traffic interruption. This multi-lane, limited-access highway serves the most industrialized

The Canso Causeway, opened in 1955, stretches across 4,500 feet of salt water to join Cape Breton Island with the Nova Scotia mainland. More than ten million tons of rock fill were required to build a roadway 217 feet deep and 80 feet wide. Until its completion travellers depended on ferries to cross the strait.

Canadian road-building in the post-war era—a far cry from pioneer statute labour! Giant power shovels and earth-moving equipment clear the way through the most rugged terrain. Base courses are laid by electronically controlled pavers, complete with sun-shades to protect the highly skilled operators. Unpaved roads, once kept in shape with home-made drags, are now maintained by power graders. But there is always a place for human muscles. The big hopper in the photograph at far right can dump more than two and a half tons of fresh concrete on the roadbed at one time. Under that pressure the rein-forcing rods are likely to shift, so men prop them in place with one of the oldest of road-building tools, the shovel.

Modern technology has revolutionized road-building. In 1921 some provinces were still using horses to power their construction equipment, but others were using mechanical loaders and light trucks as in the photograph at far left, taken in British Columbia. Today the twenty-one-ton behemoth shown in the next picture can dig a trench nineteen feet deep in a single operation.

and densely populated portion of the province, a twenty-mile-wide strip in which more than three million people live. Quebec, meanwhile, returned to the pay-as-you-drive system to finance a network of urgently needed fast superhighways fanning out from Montreal. The Quebec Autoroute Authority, formed in 1956, built Canada's first toll freeway from Montreal forty-five miles north to the year-round Laurentian playground of Ste Adèle. The six-lane, mile-a-minute autoroute replaced an outdated highway on which traffic sometimes piled up for five hours. It proved so popular, even at a charge of two and a half cents per mile, that further toll roads were built from Montreal southeastward to Sherbrooke, and from there towards the Vermont border; and from Montreal eastward along the North Shore to Berthierville, with Quebec City as the eventual goal. Not a cent of tax revenue was used to build these roads; capital costs were met by government-backed bonds issued by the Autoroute Authority, a Crown Corporation. All the tax money available was needed to maintain, improve, and extend other portions of the highway network, including a number of roads into new areas following the advance of mining and settlement.

Less spectacular but equally important, the existing road system across Canada was continuously being improved. Back lines were widened, graded, and gravelled; secondary roads and highways were straightened and freshly surfaced or paved. By the sixties Prince Edward Island, once the bitterest holdout against the motor car, boasted the largest per capita mileage of paved highway of any Canadian province —the result of devoting up to one-third of her budget for several years to roads. On the prairies, where half a century earlier road-builders had been satisfied with a sixteen-foot width, routes between major cities were being expanded to four or more traffic lanes. The first of these divided highways paralleled the old trail north from Calgary towards Edmonton. From the Atlantic to the Pacific new bridges were being built to span rivers, valleys, and harbour mouths, using the latest advances in the technology of steel and concrete.

By the mid-sixties Canada was spending well over one and a quarter

In the 1960s earth-moving equipment was outfitted with pneumatic tires eight feet and more in diameter for added power and manoeuvrability.

billion dollars each year on one type of road or another to speed her automotive traffic. The standard of highways was so high that motorists who enjoyed a challenge developed a new sport, "pike-shunning," and deliberately sought the uncrowded, unpaved back ways.

Probably nothing would surprise Canadian pioneers more than today's high-speed highways and traffic—unless it was the way the roads are built. We have come a long way since the days of axe and logging-chain. After the Second World War power machinery grew steadily bigger, more powerful, and more useful. The most noticeable change was in scrapers, used in cut-and-fill work in new construction. In 1950 the average scraper was a tractor-drawn pan which could carry some ten cubic yards of earth; within fifteen years most were self-propelled, rubber-tired behemoths that carried thirty to fifty cubic yards at a time at highway speeds. Much of the new equipment in the sixties—even the big bulldozers—rode on huge pneumatic tires instead of tracks, an innovation which increased their power, manoeuvrability, capacity, and speed. As a result the cost of moving a yard of earth remained practically the same as it had been in the 1920s.

New techniques were constantly being developed. Instead of steam rollers, heavy pneumatic devices were used to compact the road-building materials quickly and solidly, from the sub-base to the finished surface. Automatic controls were introduced to correct human error and meet increasingly high specifications for construction. On paving equipment, electronic devices were adapted to help lay surfaces smoother than man's unaided judgment could possibly achieve. Even the traditional winter shut-down gave way, as research uncovered ways of continuing work on many phases of road building throughout the months when the ground was frozen.

Road planning has been equally revolutionized. Once it was thought enough to send a surveyor ahead to mark the centre line for the soldiers

The traditional survey crew represents only part of the planning that goes into today's roads. Beyond that there is a complex programme of economic and geographical analysis—often aided by computer—which draws on the latest developments in engineering and technology.

or statute labourers. The route in those days was to a great extent dictated by nature: it avoided rocks, and when possible it went around hills to spare the horses. Today's engineers remove the rocks and hills. Their prime aim is a fast, safe highway. Gradients must still be minimized, but so must corners and curves: high-speed traffic must flow smoothly and see far ahead.

Months or years before the earth-moving equipment arrives, engineers from the planning branch of the highway department are on the scene, estimating potential demands on a projected development. By surveys they determine why people in the area travel, where most of the traffic originates and where it ends, average and peak flows at daily and hourly intervals, speed trends, and the behaviour of drivers who use the road. From this they may recommend, for example, that along most of its length a new stretch of highway between two cities can be four lanes wide; but at either end it must be expanded to six lanes to meet heavy local traffic. Will fifty per cent of the potential traffic flow directly between the two cities, and another thirty-five per cent have either its destination or its origin in one of them? Then towns and villages on the way can safely be by-passed. Here and afterwards, electronic calculators and computers are used extensively to analyse masses of figures on traffic flow, design characteristic, materials, and costs.

The actual route is determined with the aid of stereoscopic aerial photographs and a detailed ground survey which includes consideration of local soils, rock, and drainage. Large-scale drawings, showing every element in the construction, are prepared and superimposed on district maps. Safety, convenience, illumination, landscaping, and control of billboard advertising are all integrated into the overall plan. Every step in the opening of the new highway is scheduled, through the programming, design, tender call, and contract award, to performance and inspection. Highway planning and design, in short, have become part of an increasingly exact science.

The growing investment in first-class streets and highways has been accompanied by stepped-up efforts in research and development. As this was written, more than one hundred agencies across Canada were engaged in finding ways to build roads that would be better, safer, more economical, and longer lasting. Provincial, federal, and municipal governments, research councils, universities, consulting engineers, associations, and industry were involved. At the centre, encouraging and helping to co-ordinate the work, stood the Canadian Good Roads Association.

The Association was reorganized in 1950 to meet the new requirements of the post-war boom. Since its birth in 1914, membership had been restricted to governments and individuals; now it was extended to provide the widest possible representation within the highway industry, including related associations, consultants, contractors, equipment dealers, manufacturers, and truckers. A permanent secretariat was appointed under a managing director, C. W. Gilchrist. The CGRA still considered that its primary function was to advocate and encourage the development of a modern system of highways to serve Canada's transportation needs, and it steadfastly refused to change its name. But a growing proportion of its activity was devoted to finding out *how* this aim could be achieved and to encouraging a unified national approach to road problems that affected the entire country. Under a Technical Advisory Council ten committees carried out programmes in the fields of bridges and structures; construction and maintenance; economics, finance, and administration; geometric design; municipal roads and streets; pavement design and evaluation; road research correlation; soils and materials; traffic operations; and transportation planning.

In 1950 there was little first-hand, detailed, factual information about Canadian roads. In making plans the country's road builders had to depend almost entirely on data collected in the United States—and hope that it applied to Canadian conditions. It didn't always. One of the Association's most important undertakings was to initiate a nationwide Canadian road test, the most comprehensive pavement evaluation project ever attempted up to that time. Working through the CGRA, provincial highway departments began obtaining and analysing performance data each year on thousands of miles of highways. Improvements in design arising from these studies meant better roads and streets, and better value for the taxpayer's dollar.

In the first fifteen years under its new organization, the CGRA published a number of technical books to disseminate the information assembled on such subjects as frost action, winter maintenance, riverbed scour, and highway finance. It formulated nationwide standards for traffic-control devices so that drivers could travel from coast to coast and meet the same signs, signals, and pavement markings they had known at home. It produced a manual of geometric design standards for Canadian roads and streets, which was accepted by all major governments across the country. It established the Canadian Highway Safety

NORMAL CLOVERLEAF INTERCHANGE

SIMPLE DIAMOND INTERCHANGE
EXPRESSWAY WITH MINOR ROAD

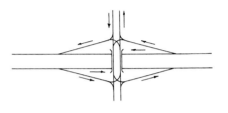

PARTIAL INTERCHANGE
EXPRESSWAY WITH MINOR ROAD

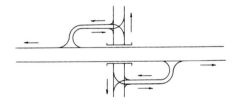

DIRECTIONAL INTERCHANGE
USED AT THE INTERSECTION
OF TWO EXPRESSWAYS

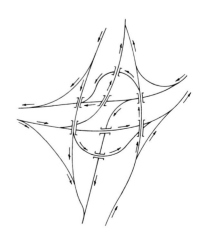

Council in 1955. It established liaison with road organizations around the world, set up a central road library, and served as a clearing house for data on research projects at home and abroad.

In 1965 the CGRA issued a list of priorities for future research. One of its top goals was a practical, on-the-spot method of quality control in road-building. If engineers could make continuous spot checks during construction they could build roads better able to stand up to traffic and frost; but the specifications and techniques they would need had yet to be devised. A more specific problem was the damage to bridges caused by de-icing agents such as rock salt: how could scientists stop the chemicals from corroding steel girders and concrete bridge decks? A third question revolved around the lower car styles of the sixties. With the driver sitting so much closer to the ground, could he still see far enough ahead for safety in rolling country, or would the hills have to be flattened?

One widespread problem was the "washboard" condition that affected dirt and gravel roads across the country. Washboard surfaces could also be found on old or poorly paved surfaces, and even on the best highways under the stress of heavy traffic. For many years the condition was believed to be due to the harmonic vibration of passing cars. Other explanations involved the soil ("too many coarse particles" according to some experts, "too many fine particles" according to others), the wind, the exhaust, the pneumatic tires, and other even less plausible

CROSS - SECTION OF BITUMINOUS PAVEMENT

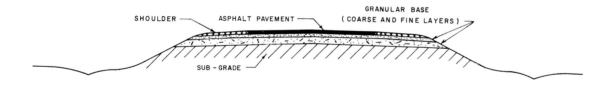

causes. Later it appeared most likely that washboarding was caused by the impact of the car wheels bouncing off random irregularities in the road surface. But whatever the reason, the shuddering effects experienced on thousands of miles of Canada's lesser roads closely resembled conditions on the old corduroy roads. The only permanent solution would seem to be better, and total, paving.

Continued study was directed towards asphalt concrete, for economic reasons the most widely used paving material in Canada. It is composed of graded aggregate mixed with a small but precise percentage of asphalt. If the proportions are wrong the surface may become dangerously slippery in wet weather, or else it might crack and wear away before its time. Highway engineers sought to learn more about asphalt's properties, knowing that if they could reduce the thickness of road surfacing by one inch they would be able to save millions of dollars over the years.

The all-too-familiar washboard surface mars an otherwise spectacular hillside drive.

While this research was going on, the many other problems of road-building were also being investigated. Bridge designs were being tested in laboratories with ingenious machinery that duplicated the pounding of traffic. Statistical relationships were being studied between accident rates and such safety features as road width, shoulders, curves, and guide rails. New paints were being developed and tested for marking centre lines.

Special attention was paid to muskeg, the gooey muck composed of water and decaying plants which covers five hundred thousand square miles of Canada. Until recently it posed an impenetrable barrier to northern expansion. Whole sections of highway were swallowed in its sponge-like morass, and in northern Manitoba it even consumed a derailed locomotive. New incentives, new vehicles, and more detailed knowledge made it possible to approach and deal with the special problems posed by vast expanses of this vegetative ooze. Road-builders who had previously shunned muskeg over three feet deep were now— of necessity—developing ways of floating roads across it. Straw, wire mesh, sand fill, piles, and even foam plastic were variously tried and

Special transporters claw their way through the bush and muskeg of the North West Territories.

used, as local conditions demanded, to provide a sufficiently stable base for a road.

Elsewhere problems were tackled with equal ingenuity. On the prairies quicklime was occasionally used to change the muddy consistency of gumbo and make a firm, concrete-like base. In Manitoba, tests were made of "insulated" road, which incorporated a four-inch layer of plastic foam in the sub-grade to slow down the penetration of frost. It seemed likely that chemicals also would be used to fight frost in the soil. New construction methods were being sought which would make less use of gravel in the sub-grade, for in some parts of the country serious inroads had already been made into available gravel deposits. One senior highway official remarked: "We are not afraid to experiment, test, or try out any new methods which might benefit our roads."

Highway engineering grew after the Second World War into a specialized profession which demanded considerable training. There were far too few people qualified to practise it. Since the early 1950s, a shortage of engineers had hampered road projects. The CGRA responded with a scholarship programme, financed by its industrial members, which would send engineers back to university for one or more years of post-graduate study or research. Then, in 1956, the University of Alberta established Canada's first post-graduate civil engineering programme concentrating on highway engineering. It included courses in highway design and construction, economics, administration, and planning, and was reinforced by basic research on fundamental problems of road construction and related problems of traffic and safety. By that time most major Canadian universities were offering one or more post-graduate courses in various phases of highway engineering.

Meanwhile highway engineers have been making more sophisticated economic studies in the course of their work. Recognizing the long-term importance of roads to the economy, provincial highway departments have begun to work closely with other branches of government to determine the impact of proposed roads on the financial well-being of the community, and to establish the delicate balance between the cost of a new project and the overall benefits accruing from it. The task is far more complex than it might appear on the surface. The true cost of a new highway involves more than the expense of obtaining the right-of-way and then building upon it. What will it cost to keep the surface of the road in good repair, to clear it of snow, to maintain the fences, signs, lighting, and ditches over the life of the new thoroughfare? What will it cost to administer it? How much will it cost to operate vehicles upon it, compared with the cost on existing routes? Vehicle costs will be affected by potential cruising speeds, grades, distances, and convenience, all of which affect gasoline consumption or the expenditure of the drivers' time. Trucks take more space and cause more damage to the highway than automobiles, other things being equal: how is this to be taken into account? Traffic volume and density can be expected to vary from hour to hour, day to day, year to year: what is the *economical* level for which the highway should be planned—and what will it be five years hence? What are the costs to the community at large? A new highway may mean loss of tax revenue; it may involve loss of amenities such as a scenic view; it will create a certain nuisance in noise and fumes for its immediate neighbours.

Against this the economist must balance the benefits of improved road transportation to private and commercial drivers, to business and industry, and to the community at large. How then can these benefits most economically be obtained? By improving existing facilities? By replacing an old road, thus eliminating faults which often can be traced back to pioneer days? By building an entirely new route and keeping the old one in operation? Then comes the question, where are the capital funds to be raised? Toll roads provide one possibility outside the public treasury. How much should be raised by gasoline taxes, vehicle registration, and other taxes imposed more or less directly upon the users of the highway? How much should be raised by general taxes upon the entire population, which includes people who benefit from the new facility only indirectly? In Canada, these questions are further

To guard the safety of the highways, police forces have taken to the air. From light planes, patrolling officers can clock speeders against carefully measured marks alongside the highways, then radio interception orders to their colleagues on the ground.

In the 1960s Canada's economy came to depend upon the efficiency of its motor transport industry.

complicated by the relationships between the federal, provincial, and municipal governments, each with its own sources of revenue, returns from which do not necessarily match its road responsibilities. (To help overcome this disparity the senior governments have on occasion offered assistance—the federal government to the provinces, the provinces to the municipalities—as we have seen.) The formulation of these questions, and their solution with the aid of advanced mathematical analysis, make up a highly specialized branch of modern economics. Without it, roads might still contribute to national development, but only accidentally; there would be a waste which no society—even one as affluent as Canada in the sixties—could afford.

For roads have become one of the most important of resources. What the waterways were in the century before Confederation, what the railways were in most of the century that followed, roads are today— the arteries which feed Canada's industry and commerce. Heavy tractor-trailer trucks rumble along the highways day and night, carrying ingots of aluminum, rolls of newsprint, tanks full of chemicals, cartons of electronic equipment, all sorts of tools and machinery, boxes of cloth goods, barrels of oysters, crates of apples, cases of canned tomatoes. Factories depend on trucks to bring the raw materials and carry away the finished goods; farmers send their milk and vegetables and livestock to market by road; builders truck in power shovels and cranes, steel girders, ready-mixed concrete, bricks and mortar; housewives watch for the delivery man from downtown department store or corner druggist. Today an entire household may be driven as a matter of course from Nova Scotia to British Columbia; or a whitefish, fresh-caught from Great Slave Lake, may be frozen, popped into a refrigerated truck, and carried over highways for 3,500 miles to grace a dinner table in New York City. Trucking has become a multi-million dollar industry, in many parts of the country the most important means of commercial transportation. The influence of roads is everywhere. Is there anything in the modern house that has not at some point travelled on tires?

Roads have played an important part in Canada's high standard of living. The industries directly associated with them—the giant automotive firms and the smaller manufacturers of automotive parts; the road-builders and maintenance crews; the men who earn their living on the roads, the truckers and taxi-drivers and travelling salesmen—form a considerable portion of the economy. About one out of every seven employed persons in Canada gains his income from highway transportation. Others are indirectly dependent upon the ease with which the goods they grow or manufacture can be shipped to the stores for sales. And the consumer benefits from lower prices and a greater variety of goods.

As the railways curtailed service in one branch line after another in the fifties and sixties in the face of trucking competition, the majority of Canadian communities came to depend entirely on road transportation, and few were without at least one highway link with the rest of the country. An exception at one time was the small town of Nakina, a railway junction point in northern Ontario, nearly one hundred miles north of Lake Superior. The hundred car owners in Nakina used the rough streets in town but could go no further than the municipal boundaries, because there were no roads beyond. Fuel oil sold in the town for $15 a barrel, compared with $9.60 in most other parts of the district; milk was twenty-eight cents a quart. It didn't look as if the province would be able to build a road to Nakina for five or ten years; meanwhile, if one of the residents wanted to drive to the Lakehead or southern Ontario, it cost him $40 to freight his car one way to Long Lac, on Highway 11. In 1953 the men of Nakina took matters into their own hands. Using volunteer labour and contributed equipment, they started building their own road through forty-five miles of dense bush to Geraldton. The route they began is now Ontario Highway 584.

While the story of Nakina is a good example of the importance of highways, surely the most striking fact in it is that a hundred people in this small town should own cars purely for local travel. Nor was this case unique. Automobiles were in great demand in isolated villages of Newfoundland before the roads were extended to reach them, and in the iron-ore boom-town of Schefferville, Quebec, hundreds of miles north of any highway. Canadians naturally depend on cars to carry them to the day's work or an evening out, to school and church, for shopping and visiting friends.

Canadians in fact are among the most travelling people in the world —and growing more so every year. In the sixties they drove their cars and trucks fifty billion miles each year over highways, roads, and streets. That was nearly twice the mileage travelled a decade earlier. It was the equivalent of ten million individual trips along the Trans-Canada Highway from St. John's, Newfoundland, to Victoria, B.C.; or, put another way, an average of nearly 2,500 travel miles per person.

Road-builders today face a deadly challenge: How can the highway network be improved to keep pace with the new cars?

The roads carried about eighty-five per cent of all Canada's traffic, measured in passenger-miles. The majority of trips were for business; others were for daily chores such as shopping; but a good number (one-fifth in some areas) were for pleasure. It was estimated that ninety per cent of the nation's vacation travel was by bus or car—in the latter case often with a house-trailer behind, or with a tent and camping gear. Canadians were seeing more of their own country, and it was no longer rare to meet a family which had driven to every one of its major regions.

Tourism was not a new industry in Canada. Back in the twenties Quebec and British Columbia had been among the first to recognize the economic significance of their scenery—the one natural resource which can be exploited indefinitely without running out. But in the years after the Second World War tourism blossomed at a great new rate, encouraged by faster, bigger cars, shorter work weeks, longer vacations,

The Gaspé coast is a favourite tourist area, thanks in large part to the highway which circles the rocky peninsula.

and general prosperity. Drama and music festivals, winter carnivals, and other seasonal activities sprang up to complement the beauty spots and sports which were the traditional attractions. Every part of Canada had its visitors from other provinces and from the United States, bringing with them dollars to augment the local economy (and in the case of American tourists, to help redress Canada's chronic trade imbalance with its southern neighbour). Perhaps the greatest single factor contributing to this development was the rapid improvement in the highway system. In 1955 more than fifty per cent of tourists complained about bad roads or traffic congestion. Within a decade roads earned more compliments than complaints. And where better roads led, tourists were quick to follow.

In the fall of 1960 a section of the Trans-Canada Highway was opened along the rocky east shore of Lake Superior. Until then this thinly-settled part of Canada had seen only a trickle of road traffic. The next summer up to three thousand cars a day drove through. The

Everything travels by road now, including boats. This one sprang a leak —in a tire.

economy boomed. At little Wawa, for example, three new motels, several cabin camps, six new service stations, a general store, a drug store, two gift shops, and a bank all opened in the first year of the new highway; township assessment rose more than a million dollars.

Across the country similar stories were being told. Hundreds of millions of dollars have followed the highways, to be invested in hotels and motels, cabins and campsites, resort areas and auditoriums, provincial and national parks. There is even one stretch of road which is a famous tourist attraction in its own right—the so-called Magnetic Hill near Moncton, New Brunswick. Motorists drive to its foot, stop their cars, turn off the motors—and coast eerily but undeniably to its top. In truth it's an optical illusion: there is no magnetic lode pulling the vehicle against the force of gravity; the apparent uphill slope is only a less steep continuation of a general downgrade. The "magnetism" extends only to the visitors, who come there from around the world.

Far beyond tourism and trucking, the impact of improved highways is felt in almost every area of Canadian society. As the last few gaps were being closed in the giant Macdonald-Cartier Freeway, a distinguished geographer, Professor E. G. Pleva, called it "the most important single development changing the economic and social patterns of Ontario." The highway was probably the longest freeway under one jurisdiction in North America. Along its route new industries had sprung up, lured by the ease with which workers could reach their jobs and materials could be shipped to market. Towns which it skirted changed from quiet rural communities to manufacturing centres; and small cities which had built up their industry in the past because they were located on the older highways now lost it as companies moved their plants closer to the new main artery. The freeway altered the growth patterns of Toronto, London, Windsor, and other large centres of population. Their metropolitan areas sprawled outwards faster and farther than anyone had imagined, as suburbs sprang up along the superhighway and its connecting roads. These new satellite communities were planned for a society which did not have to depend on its legs. The houses had plenty of green space around them. The people who lived in them drove on the freeway to the city to work, and to the country for recreation. They shopped along its route at giant shopping plazas—a post-war marketing development entirely dependent upon the automobile. They used it to go to schools and hospitals. They followed it to keep up friendships with families who lived miles away. The Macdonald-Cartier Freeway changed the economy of southern Ontario and the social, working, and spending habits of its people. And so in its own area, to a greater or lesser exent, has every other road that Canadians have built.

Nowhere have the far-reaching effects of improved road conditions been more apparent than in Canada's youngest province. When Newfoundland became part of Canada in 1949, road facilities on the island were still as primitive as they had been on the mainland in 1914. The rapid change that followed Confederation was described at the CGRA's fiftieth anniversary by its then president, Hon. F. W. Rowe, Minister of Highways for Newfoundland:

The majority of Newfoundland's 1,300 communities had no road connections of any kind in 1949. These communities, which were tied together for a few months during the summer by narrow gravel roads, had to resign themselves to six months of winter isolation. But a revolution has taken place.

In that 15-year period we have built 2,100 miles of new road, rebuilt 2,500 miles of old road, paved 600 miles, completed 80 per cent of a 580-mile Trans-Canada Highway, and built 225 major bridges at a cost of $20,000,000. In all, our total expenditure on roads in that period has reached the staggering sum of $240,000,000. But the real significance of this achievement cannot be grasped by citing miles, or dollars. It lies, rather, in the impact of this investment on human beings.

All but about one hundred settlements in Newfoundland are now linked together. To put it another way, more than one thousand communities have been relieved of the historic curse of isolation. Hospitals and doctors are now able to minister efficiently to the medical needs of the people. The great majority of the people are within easy access of social welfare and other government services. Most important of all, educational authorities have been able to centralize and consolidate educational facilities in such a way that many of the one-

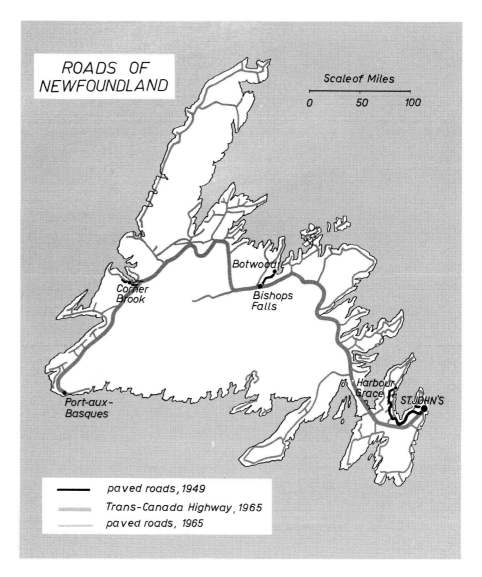

ROADS OF NEWFOUNDLAND

Scale of Miles

0   50   100

Botwood

Corner Brook

Bishops Falls

Port-aux-Basques

Harbour Grace

ST. JOHN'S

——— paved roads, 1949
——— Trans-Canada Highway, 1965
·········· paved roads, 1965

and two-room schools have now disappeared and the majority of high-school students are now transported daily to a network of central and regional high schools.

One example will serve to illustrate the revolutionary impact of this development. Even as late as 1955, the average child born in a community with a one-room school had one chance out of 700 of attaining a Grade XI matriculation standard. Today, the average child in such a community has the same chance as if he were living in St. John's or Montreal. Largely because of roads, the Province of Newfoundland has undergone a revolution—economical, sociological, educational, and psychological.

In Newfoundland, condensed in fifteen years—in a fast-motion film, as it were—we may see what has happened across Canada in the twentieth century.

*Building the Bronte Bridge west of Toronto on the Queen Elizabeth Way,
September 1936. Canada's first superhighway leaped across the ravine crossed
so painfully 136 years earlier by Asa Danforth.*

BRIDGES III

After the Second World War new steel spans pushed out across space in every part of the country. *Far left: Over the Little Pic River, twenty miles west of Marathon on the north shore of Lake Superior, for the Trans-Canada Highway. Left: Construction of a new bridge over the St. François River, Drummond County, Quebec. Right: Causeway-bridge across Great Bras d'Or Lake in Cape Breton Island, another link in the Trans-Canada Highway.*

A gallery of Canadian bridges. Clockwise from top left: Burlington Skyway at the western end of Lake Ontario; Petrofka Bridge over the North Saskatchewan River; Nine-Mile Canyon Bridge, carrying the Trans-Canada Highway through the old Cariboo gold rush region; White River Bridge, 1,169 miles up the Alaska Highway from Dawson Creek; in the cities (Vancouver in this instance) modern bridges leap over buildings as well as water to speed urban traffic; the Curtis Causeway and Bridge at Dildo Run, between New World Island in Notre Dame Bay and the Newfoundland mainland.

Only a few of Canada's major bridges can be shown on these pages. Some of the missing ones are tourist landmarks almost as well known as the Mounties' red coats—the Rainbow Bridge at Niagara Falls, for example, or the Lions Gate Bridge across the mouth of Vancouver Harbour. Since nearly forty-four per cent of Canada's southern border is water, bridges play an important part in joining our nation for travel and commerce with the United States. Other bridges represent considerable technological achievement. On the east coast the Angus L. Macdonald suspension bridge, almost a mile long, links the cities of Halifax and Dartmouth. At Arvida, Quebec, the world's first aluminum bridge spans the Saguenay River; it is 290 feet long and weighs about half what it would if it were made of steel. At Port Mann, British Columbia, traffic flows without interruption across the Fraser River on a 6,870-foot structure opened in 1964. For the driver bridges afford comfort, convenience, and an element of artistry in Canada's landscape.

The Trans-Canada Highway winds through Glacier National Park, British Columbia.

15

A Mari Usque Ad Mare

IN 1870 BRITISH COLUMBIA was still a separate British colony, undecided whether or not it would join the young nation of Canada. When its representatives went to Ottawa to negotiate its confederation, one of their main concerns was a land link with the other provinces. Ultimately they wanted a railway, and they asked that a survey begin at once; but because this would be a stupendous undertaking they were prepared to accept a temporary expedient. They would be satisfied with a wagon road from Winnipeg to Vancouver, to be completed in three years. For reasons that are still not clear, the federal government shrugged off this half-measure. It promised to begin a railway within two years, and to have it operating within ten. That decision gave Canada one of its great legends, the building of the Canadian Pacific, but it doomed the chances for a connecting road, which was no longer necessary. The better part of a century would pass before Canada would achieve a continuous highway from coast to coast.

A junior version of the Trans-Canada Highway had existed as early as 1827. It was then possible—just possible, if one was determined and durable enough—to travel along the pioneer roads from Halifax through Nova Scotia and New Brunswick, across Lower and Upper Canada, to the western fringe of settlement at Amherstburg, near Windsor. By the 1870s there were roads further west: the Dawson Road had been cleared from the Lakehead to the Red River Valley, and from there rough trails cut across the grasslands to the Rockies. From the Pacific, the Cariboo Road carried stage coaches eastward into the interior. Only two long gaps remained to complete the transcontinental route, but both were formidable. One was the long stretch through the rock and muskeg of the Canadian Shield across the top of Lake Superior; the other was the shorter but even more intimidating mountain passage over the Continental Divide.

In the heyday of the iron horse nothing more was said about a road from coast to coast. The project slumbered, at last to be awakened by the honking and backfiring of the horseless carriage. In 1910 a group of western pioneer motorists founded the Canadian Highway Association to promote construction of a road across the country. The Association held two conventions. In 1912 it was entrusted by A. E. Todd, of Victoria, British Columbia, with a gold medal to be awarded to the first person driving a motor vehicle along an all-Canadian overland route from Halifax to the Pacific. A few months later the Association died. Its goal was too distant to sustain the necessary interest.

Just how far off a trans-Canada highway really was in 1912 can be discovered in the story of Thomas W. Wilby, an intrepid Englishman who sought adventure and then wrote about it. His book, *A Motor Tour through Canada*, published in 1914, makes good reading. As he explains, the trip seemed straightforward enough when it was being planned in England. There was only one way across: from Halifax to Saint John; north to Rivière du Loup; then up the St. Lawrence to Quebec and Montreal, and from there to Ottawa; westward to North Bay and Sudbury; alongside the Canadian Pacific Railway to Winnipeg; across the prairies in a more or less straight line by way of Brandon, Regina, Moose Jaw, Medicine Hat, Lethbridge, and Fort Macleod to the Crowsnest Pass. After that, anyone would be able to tell him how to get to Vancouver!

On August 27 all was in readiness. Wilby's big touring car stood with its wheels in the Atlantic Ocean. Spare tires hung jauntily at the back; a polished metal trunk protected the suitcases in the tonneau; two long boxes on the running boards held spare parts and the emergency gasoline cans; fully loaded, the machine weighed two tons. Beside Wilby in the front seat sat F. V. Haney, his Canadian mechanic. They checked the brake pedals, the gear lever, and the foot-worked horn. Wilby dismounted to fill a bottle with water from the eastern ocean. Then the motor roared and the trip bravely began.

Doubts must have begun soon afterwards. The road to Truro, through dark woods, grew narrow and winding. After weeks of steady rain its clay base was sodden with water, the consistency of a batter pudding, and the surface sprinkling of gravel or sand was little help. To advance required "a good deal of determination and petrol." But so far Wilby had met nothing exceptional. His route through the Maritimes and Quebec was relatively easy—by the standards of the day. He had his small adventures, to be sure: travel along a grassy cart-path marked only by wheel-ruts and horses' hoofs; a wrong turn because of undersized directions on the signboard at a fork; frequent battles to push the car through herds of cattle on the roadbed. Yet the journey also had the air of a triumphal procession. Everywhere along the route to Vancouver Wilby was received and helped by mayors and boards of trades, motor clubs and highway associations. People volunteered as pilots to guide him on the next step of the journey. Every day's run—generally 150 miles or less—was reported in the newspapers.

The first check came at Ottawa. It was impossible to cut directly across to North Bay: Wilby would have to dip down to Toronto and back up again in a long U. It took two days to drive to Toronto. On particularly good sections there was usually a toll gate, and a charge of as much as twenty-five cents. "It would only have been fair," Wilby wrote, "had the authorities in their turn paid us whenever we had to travel a bad stretch of highway." Once the car was almost lost in a muddy ford, but they got it out with the aid of mudhooks—"a diabolical contrivance attached to the rear wheels whereby the machine jerked itself violently to safety."

From Toronto they followed Yonge Street northward to Lake Simcoe. Beyond that began the Canadian Shield. The car pushed its way through overgrowths of hemlock and birch, dodged boulders, and finally made its way onto the sandy main street of Gravenhurst. And the next day Wilby's troubles really began.

No car could preserve its sweetness of temper on these Mid-Ontarian grades. Our own particular machine showed ill-humour whenever her front paws struck the steep pitches. She growled and gave out little snappy short barks, forcing us to get down and adjust the carburettor to the new altitudes. Forewarned, we had armed ourselves with a handy compression pump directly attached to the petrol tank under the seat. Compression meant more climbing power, and affected her as oxygen does the athlete. . . . As we rocked and plunged along the stony path a human settlement appeared in a dip below, Novar! A ribbon of road ran to meet it from somewhere in the rear. . . . The path was crowded

Thomas Wilby followed trails "straight as an arrow across the illimitable prairie" during his epic cross-Canada journey in 1912.

with stumps of trees. . . . We struck a wallow of deep tractionless stuff between the innocently tempting heap of sand at the foot and on the crest. Wheels dropped to axle level and motion to stagnation. There was nothing to do but to try and rush the ascent. But the hill was too wary to be caught off its guard. We retired ignominiously to the foot again. . . . We were caught like a fly in the tangle of a spider's web. . . . The sand flies from her rear as from the mighty back kick of an ostrich leg. But her spirit is broken at last. The wheels spin on the same spot in frantic, static dynamics. . . . She admits defeat. . . . Only horses can now bring life to the useless, motionless mechanism.

The driving shaft had broken. Wilby marked time until it was replaced; at last the trip resumed, but now there was scarcely a road of any account to follow. The printed guides he had been given were useless. Local settlers directed the travellers through mud and sand and swamp, or over stretches of corduroy. Tires burst. Fresh hills rose in the way—each a challenge. At one, Wilby shook his head in despair. It was impossible to rush the slope, for the corduroy at its foot would have broken the springs and landed the car upside down in the roadside bush. He gave the big motor every ounce of power and hoped. The car stopped dead less than half way up, and began sliding backwards. A hummock held it long enough for them to jump out and block the wheels. Then began a long battle. They tried compression on the motor; they jury-rigged a windlass with birch stumps; they jacked up the wheels and filled the ruts beneath them with stones; and finally the battered vehicle reached the top. But the new shaft was twisted, and they could only crawl in low gear. It had taken them five days to travel two hundred miles, and—though they could not know it—they would still have to rebuild a missing bridge before reaching North Bay.

Here Wilby finally had to admit defeat. There was no uninterrupted road west of North Bay. Reluctantly, he proceeded to the railway station. The next eighty miles were covered behind a snorting locomotive, with the automobile locked ignominiously inside a freight car. Then came a brief attempt at a newly-opened government road to Sault Ste Marie. It nearly ended in disaster. Wilby finally took a tugboat to the Sault, and a larger boat across Lake Superior. At the Lakehead he capitulated again to the railway, and freighted his car the rest of the way to Manitoba.

In Winnipeg he found that nature was still throwing up obstacles. Rain had fallen all summer, and throughout the province automobiles were caught helplessly in black gumbo: there was nothing to do but wait till the sun broke through. Even then, progress along the spongy roadbed was slow and laborious; at times the car sank into the mud up to its hubs. But the path was straight most of the way at least, and where it left the gumbo and ran across virgin prairie Wilby found the finest natural road he had ever met—"a treeless lonely expanse, cut by wheels of waggons, scarcely furrowed by rains, worked here and there into lumpy incrustations by the feet of cattle."

West of Brandon there was not a vestige of constructed highway, only a path that wandered and listed over broken ground, through wild tangles of bush and swamp, past alkali outcroppings. Alone, they would have been lost countless times, but the local school inspector who was their guide knew every corkscrew turn and zig-zag. He led

them safely across the Saskatchewan border to Moosomin, where they picked up another guide. Once they missed a fork on the rolling bush trail and found themselves enveloped in mud and swamp; other times they crossed ploughed land or invaded farmyard, but at last they reached Regina. The next leg, to Moose Jaw, carried them over typical open prairie. Two other cars led the way.

We had neither sign-posts, milestones, nor other landmarks. . . . The road soon showed signs of fickleness. It varied at every half mile. It lost its fence, then regained it. It lost its wheel ruts, then picked them up again. Then the prairie grasses, merely held back by a strip of wire fence, boldly crept into the roadway, cheerfully sprouted all over it. . . . A microscope, then, would have failed to find the road, and the barbed wire and sun became our sole guide in endeavouring to "hit" the government trail. . . . Here and there the long narrow parallel trails of the bygone buffalo paths, worn bare in the grasses along which the animals used to march in single file, streaked their way across the landscape.

Where there were no homesteaders the trail failed; where the land had been taken up the road began again, but then there were gates to open and shut, fences to dodge. In places the roads were alive with little yellow gophers who waited until the wheels were almost upon them before turning tail.

And then, near the Alberta border, the prairie was left behind and they entered ranching country where the trail was "smooth as a tennis lawn." The mountains came closer and closer. The road twisted and coiled upward. Grades proved most deceptive: seemingly mere slopes, they called for every ounce of power; Wilby would have to run alongside the car, helping to push it up the hills. At the Crowsnest Pass they crossed into British Columbia and pressed on to Cranbrook. Again Wilby was advised to take to the train. This time he refused.

Between Cranbrook and Nelson, the next point on Wilby's planned itinerary, lay a range of mountains crossed only by a narrow trail impassable for automobiles. His only hope—and it was a slim one— was to drive south instead of west, almost to the United States border, part of the way through a swamp that had never before been attempted by motor car. At Yahk he could turn north again, if he was prepared to risk travelling on the railway ties over fourteen miles of dangerous loop track. After that lay a very steep and narrow mountain trail over the Goat River Gorge to Creston; there, if he got so far, he would probably find guides to lead him through the flats to a steamer, which would carry him up Kootenay Lake to Nelson.

South from Cranbrook, on the first leg of the trip, one of the local drivers led the way in his own car. Wilby's heavily-loaded transcontinental tourer followed with difficulty. By the time they reached the swamp, night was already falling. They lit their acetylene headlights and warily edged forward. At times both vehicles sank in the slough up to their hubs, listing heavily, grinding and ploughing, while the engines roared and the wheels shot inky water out over the men. The last few miles to Yahk they had to be pulled by horses.

Then came the nerve-wracking, spine-jolting trip over the railway tracks. At any moment a train might come hurtling round a bend and fling them over the embankment. With muscles tensed to leap to safety, they strained their eyes for the gleam of a locomotive headlight.

Where there were no roads Wilby either hoisted his auto onto a railway flatcar or—for one breathtaking passage—bumped over the ties.

Meanwhile, their teeth were almost rattled out of their heads as the car wheels bounced from tie to tie. The track ran sharply downhill. The curves grew sharper and shorter. The wheels caught in the frogs of a switch: hurriedly they jacked the car up and got it clear. The track spikes cut the rubber tires to ribbons. But fortune was with them—no train appeared, and at last they reached a lonely, darkened station in the forest, where they could pick up a road to Creston.

Worse country was to come. At Nelson, Wilby had to take a train again for twenty miles. Occasionally he had to make his own path. Horses and rafts were requisitioned to help through mud and over rivers. The car rattled down the Thompson valley and the old gold route along the Fraser, through tortuous country where rock slides were a constant danger. On the Cariboo Road it edged past strings of nervous pack horses and an old-fashioned wagon train. Once the headlights failed: an extra driver came to the rescue, lying on the front fender for ten miles, holding an oil lamp close to the road's outer edge.

But now the long journey was almost over. They rolled through Lytton and Yale and Hope. Near Chilliwack the car was met by a welcoming escort. At Vancouver, Wilby delivered to the mayor a letter from the mayor of Halifax. Then, determined not to stop until he had travelled as far west in Canada as he could, he put the car on the steamer to Vancouver Island. At Victoria, escorted by city officials and members of the Victoria Automobile Association, the road-weary vehicle glided along the sandy beach and dipped its tires into the ocean. Fifty-two days before, Wilby had filled a flask with water from the Atlantic. Now the eastern ocean was emptied into the Pacific—a solemn libation to the future transcontinental traveller.

In nearby Alberni stood a post with the words "Canadian Highway" painted upon it. An arrow pointed due east to Halifax. This, in the fall of 1912, was the first signpost on the projected Trans-Canada Highway. It had been raised hopefully a few months earlier by public-spirited men from Victoria, Vancouver, and Nanaimo as the symbol of a national need. Before long they were joined by other groups. The Canadian Automobile Association was formed in Toronto in December 1913; the following year, the Canadian Good Roads Association held its inaugural meeting, billed as "The First Canadian and International Good Roads Congress." Both associations asked the federal parliament to help the provinces financially with road-building; both sought, among other things, a uniform approach to road-building among the provinces. The First World War slowed down the movement almost at its inception, however, and in the post-war years a trans-Canada highway was rather lost sight of among the more immediate needs for market and suburban routes.

Still the pressure for a national road continued. In 1919 it received its first major encouragement. The Canada Highway Act, which provided $20 million in federal assistance to provincial road departments, was passed partly with the cross-country project in mind. It called on the provinces to plan their new developments so as to form a trunk system. The maps they submitted with their applications for funds showed, as a result, an embryonic trans-Canada highway—though one which

changed at each provincial boundary and was broken by a long gap north of Lake Superior. A courageous photographer, Ed Flickenger, tested the results in 1925 in a Model T Ford. He found he could follow the highway system for 4,000 miles—some of the way on asphalt or concrete surface, much of it on washboarded gravel, and a good deal of it over plain dirt. For 850 otherwise impassable miles in northern Ontario and the Rockies, he fitted his car with flanged wheels and drove along the railway tracks.

One of the most vigorous supporters of the transcontinental route at this time was A. W. Campbell, who had been Ontario's first Instructor in Roadmaking. Now he was in Ottawa as Chief Commissioner of Highways for the federal government. Another supporter was Dr. Perry Doolittle, whose name belied his energetic cross-country campaigning. As president of the CAA from 1920 to 1930 he drove repeatedly over every mile of the trans-Canada route open to traffic, stopping at each major centre to speak in favour of its completion. George A. McNamee, for thirty-six years secretary-treasurer of the CGRA; A. C. Emmett, secretary-manager of the Manitoba Motor League and credited with devising the numbering system used on Canadian highways, and J. A. Duchastel, fourth president of the CGRA, were also outstanding in the crusade.

During the 1930s the provinces received another $20 million from Ottawa for trunk highway development. At the same time they were obtaining more revenue from automobile registration and gasoline taxes; and during the Depression highway building provided useful relief work for men who would otherwise have been unemployed. The year 1940 saw two notable achievements. In British Columbia the Big Bend Highway was opened through the mountains from Golden to Revelstoke—a 190-mile gravelled stretch which took eleven years to build. Its path followed a long northern loop in the Columbia River between the two towns, which are only fifty-seven miles apart as the crow

Ontario brought the Trans-Canada Highway one step closer to reality in 1932 with the completion of a route from Kenora west to the Manitoba border.

In some areas of Canada the only way to travel overland during much of the twenties was by rail, with flanged wheels fitted on a truck or automobile. Right: Some drivers carried their own turntables so they could change direction at the end of the run. After use the turntable would go back on top of the auto.

Above: In the Marathon area of northern Ontario the new highway was built through land never before opened to wheeled traffic.

Below: The nation's new "main street" is carried over the tracks at Moose Jaw, Saskatchewan. The dangers of level crossings were eliminated wherever possible.

Blasting signalled the start of construction in Banff National Park as the dream of a cross-Canada highway approached reality in the fifties.

flies. Snowfall was so heavy that the road could not be kept open in winter; and even in summer, driving along the Big Bend was a hair-raising experience. Those who did it talked for years afterwards of passing rusting wrecks of other cars, fording mountain run-offs, dodging rocks from slides, bouncing over potholes and washboard, edging along cliff sides, and inching up steep grades. But tough as it was, the Big Bend was the final link west of the Great Lakes in a true Trans-Canada Highway.

The second event in 1940 was the completion in northern Ontario of the 106-mile section from Nipigon northeast to Geraldton. East of that there was still no highway. To reach Hearst, about 150 miles from Geraldton, the shortest motor route was 2,070 miles around the Great Lakes. Yet this last gap was succumbing to the builders. Two-thirds of the way was already cleared, though not yet graded, and when it was finished Canada would have truly achieved a national road.

For a second time progress was slowed by world war; but with the return of peace in 1945 a Dominion-Provincial Conference recognized

highways as a matter of national concern. The next year the Todd gold medal finally found an owner. Two Canadians, Brigadier R. A. Macfarlane, D.S.O., and Kenneth MacGillivray, made the first trip across the country without resort to railway tracks. It took them nine days to drive their 1946 Chevrolet from the Atlantic at Louisburg to the Pacific at Victoria. This time the 4,743 miles which had played havoc with Wilby's car were conquered at the cost of only four flat tires.

In 1949 the curtain opened on the final act of the drama. Parliament passed the Trans-Canada Highway Act, which provided federal assistance in building a hard-surfaced, all-weather road from coast to coast. Ottawa would pay fifty per cent of the estimated cost of $300 million, and up to fifty per cent of prior construction since 1928. Along its entire length the highway would be at least two lanes and twenty-four feet wide, with ten-foot gravel shoulders, maximum grades of six per cent, curves of no more than three degrees, and a minimum stopping sight distance of six hundred feet except in difficult country such as the mountains. The provinces would be responsible for choosing the

At Rogers Pass high in the mountains the builders of the Trans-Canada Highway took extraordinary precautions against avalanches. Roaring down the mountainsides, the massive snow slides (top left) would carry away any driver unfortunate enough to be on the route at that moment, and leave the road blocked for days. Over the highway itself,

artificial tunnels (top right) were built of concrete and steel. On the slopes above, earth mounds (bottom left) were erected along known avalanche routes to break up the falls before they could reach road level. To safeguard the construction crews, army howitzers were fired to touch off any potential avalanches before work began.

shortest practical route, and for maintenance. The highway was to be completed by 1956.

Work began the following summer. The job proved more difficult and expensive than anyone had predicted. As time ran out the completion date was set forward to 1960; the federal government, moreover, aware that some stretches were costing up to a million dollars a mile, agreed to pay ninety per cent of the cost of building ten per cent of the mileage in each province. In 1957 Saskatchewan announced that it had completed paving the 406 miles within its boundaries; it was the first province to do so. By mid-1960 good progress had been made—three miles out of every five along the route were paved to Trans-Canada Highway specifications—but obviously still more time was needed. The cost by then had crept to more than $586 million. By 1965 it had approached one billion dollars, and construction crews were working feverishly to have the last few gaps in the paving closed by the end of the Centennial year.

The highway builders at last had subdued Canada's most rugged countryside. East and north of Lake Superior, from Agawa Bay to Marathon, they had forced a first-class highway through land that had never before been penetrated by any kind of road. That work took four years. To some points on the route, men and equipment were carried by barge and boat; to others they were flown in by airplane or helicopter. The workmen blasted cuts three hundred to six hundred feet long and up to seventy-five feet deep through solid rock. They moved nearly 1,300,000 cubic yards of rock and dirt, and built twenty-five major bridges, one of them 590 feet long. The final result was an asphalt highway with easy curves and grades through country which previously had been closed to anything but foot travel.

In the west the Cariboo Road was built again, this time with major new bridges and reinforced concrete retaining walls. Transformed into a sixty-mile-per-hour highway even through the Fraser Canyon, it continued to be British Columbia's main north-south traffic artery. Further in the interior the Trans-Canada Highway was carried 4,400 feet above sea level on a new, more direct link between Revelstoke and Golden. One hundred miles shorter than the Big Bend highway, hard-surfaced, designed with gentle curves and good grades, this stretch is probably the most spectacular accomplishment of the entire transcontinental road. It crosses the Selkirk Range at Rogers Pass—"death pass" in the old days, where steep slopes and heavy snowfall combined to make rock slides and avalanches common. The CPR line used to go this way, but after 236 workmen were killed in thirty years the railway decided to cut its losses: it built a five-mile tunnel to carry its trains in safety. The road-builders have erected a different kind of tunnel. Rugged steel-and-concrete walls and roofs cover the highway in areas where rock slides and snow slides have been known to occur. These structures can withstand 1,200 pounds of pressure per square foot. On the slopes above, mounds known as dragon teeth have been made to divert and break up the snow slides which would carry away anything in their path and block the highway to traffic for days. Even as they were building the protective devices, construction crews had to dodge falling rocks. When they finished, the Trans-Canada Highway was considered officially complete. The formal opening took place at Rogers Pass on July 30, 1962.

The 4,860-mile Trans-Canada Highway is the longest national highway in the world. It begins at St. John's, Newfoundland, where Devon fishermen founded a settlement early in the sixteenth century and Jacques Cartier stopped in 1542. From there it meanders 565 miles in a horse-shoe curve north to Gander and on to Grand Falls and Cornerbrook, ending at Port-aux-Basques in the southwest corner of the island. Here the motorist must take the ferry to Sydney, Nova Scotia. The highway extends westward past Bras d'Or Lake, where the first powered flight in the British Empire took place. Near Pictou there is another ferry ride to Prince Edward Island, which has the shortest stretch of the highway of any province—only seventy-one miles—and one of the most historic capitals, for it was at Charlottetown that the principles of Confederation were hammered out. Then back to the mainland, to Moncton and Fredericton, and on up the old Temiscouata Trail to Rivière du Loup. No need to portage any longer! Along the south shore of the St. Lawrence the highway runs to Quebec, carries on to Montreal, and up the old fur-trade route along the Ottawa River to the nation's capital. Westward then to North Bay, past the nickel mines of the Sudbury region, and on to the great locks at Sault Ste Marie by which the lake ships pass from Superior to Huron. Here the highway enters the newly-conquered territory along Lake Superior, then through Wawa and Marathon and Schreiber, to the twin cities of Port Arthur and

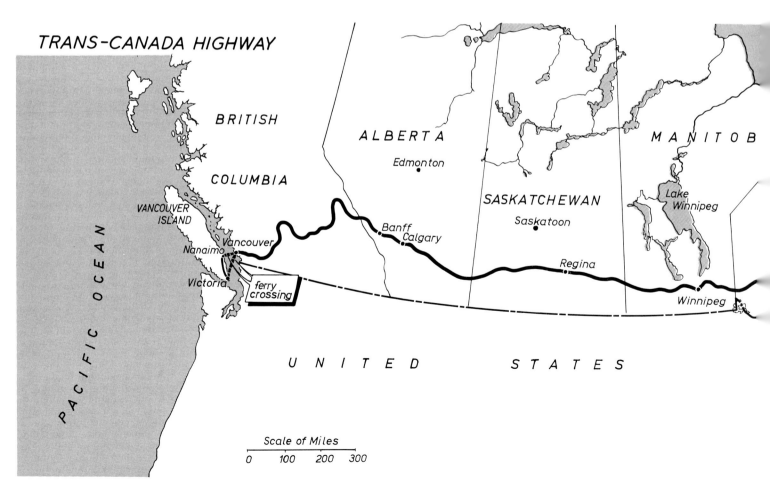

TRANS-CANADA HIGHWAY

Fort William. From the Lakehead it runs north of the old Dawson Road to Kenora, and on to Winnipeg where the Red River carts used to begin their journeys across the prairies. Now western travel is vastly different. The Trans-Canada Highway runs straight and smooth—and dust-free—to Regina, which used to be called Pile o' Bones because of the buffalo skeletons heaped there, and on to Moose Jaw; past field after field of wheat; through Medicine Hat and Calgary, where the oil derricks rise; and then into the mountains—the long climb past Banff to the Great Divide; the tamed Rogers Pass; through the valleys to Kamloops and Cache Creek, and down the old Cariboo Road where stagecoaches used to travel with their loads of gold and fortune-hunters; past Spuzzum and Hope to Vancouver. Again a ferry is necessary, for the road ends at Victoria as it began—on an island.

And the road is travelled. Since the mid-fifties Canadians have been driving inside their own country as they never did before. Everywhere along the Trans-Canada Highway new tourist accommodation sprang up to meet the needs of the new mobility. For the first time a factory worker in central Canada could take his family on an economical two-week vacation to the wilderness above the Upper Lakes, or to the Maritimes for a view of the ocean. Even more significantly—for most of the eastern route, in one shape or another, was well established—he could easily head west by car within his own country's borders. The highway surmounted two of the greatest barriers nature has flung across our nation. The Canadian Shield is no longer an obstacle to be by-passed by a southern detour. Once the highway was completed, tourists surged

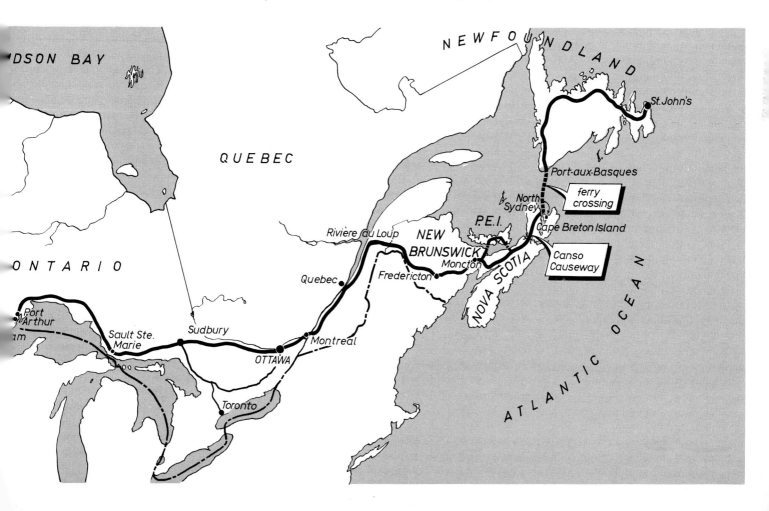

over Lake Superior into Manitoba in such numbers that in one eighteen-month period alone $1.5 million was spent building motels east of Winnipeg to accommodate them. And the mountains have also surrendered. With the completion of the Rogers Pass road, Calgary became the de facto capital of the British Columbia interior. Families drove down from the mountains to shop in its stores and enjoy its attractions; Calgarians gave up their traditional holiday resorts on the lakes of Idaho and Montana, and played instead in the Okanagan Valley, only a matter of hours away by the new good roads. If they wanted to carry on to the Pacific, Vancouver was only a twenty-four-hour trip.

The full impact of the new national road was only beginning to be felt as Canada approached her hundredth birthday. For most of the time since 1867 it had been a commonplace to say that the country was held together by ribbons of steel—the tracks of the transcontinental railways. Canada could start her second century with a new tie, a ribbon this time of concrete and asphalt, a unity based on road and wheel.

For Further Reading and Research

SOURCE BOOKS relating to the history of roads in Canada are not generally accessible. Materials on the subject are widespread, but relatively few of them can be obtained at any one location. In general the researcher in the history of transportation will wish to consult museums, archives, libraries, newspaper files, galleries, private picture collections, highway departments, road and trucking associations, and the files or archives of private companies engaged in transport. The Public Archives of Canada and the archives of every province are invaluable sources of national and local historical material; so are the provincial departments of highways or of public works: the Department of Public Works in Fredericton contains an excellent collection of photographs of roads, bridges, and ferries from various periods; the Department of Public Works in Winnipeg has extensive material on roads and trails, dealing largely with Manitoba's historic trails and the manner in which they evolved into modern roads. Additional special sources are the Quebec Museum, Quebec, for unpublished materials and illustrations relating to the history of transportation in that province; the John Ross Robertson Collection, Toronto, for historical illustrations pertaining to Canada; the Sigmund Samuel Canadiana Gallery (Royal Ontario Museum, University of Toronto), the Public Archives of Canada, and the National Gallery of Canada, for paintings and drawings of early Canadian life and conditions; the National Film Board, Ottawa, for a great variety of photographs of construction, roads, and vehicles across Canada; the Glenbow Foundation, Calgary, for unpublished materials on prairie roads and trails; the City Archives, Vancouver, for historical photographs dealing with roads in the west; and the libraries of the British Museum and Cambridge University for rare books on travel in Canada.

The following entries have been selected to provide a broad basis for study of the history of roads and transportation. The books which are of special interest to the student or general reader are marked with an asterisk. In a few cases titles have been abbreviated.

I UNPUBLISHED DOCUMENTS

"A Brief Outline of the History of Roads in Newfoundland, with Statistics and Expenditures," Department of Highways, Newfoundland.

"Agreement between the Cobourg and Rice Lake Plank Road and Ferry Company and Certain Millers," signed October 5, 1847, by William Weller, president of the company, and three saw-millers who supplied planks for the Road. Reference Department, Toronto Public Library.

BERGERON, ARTHUR
"Extraits du mémoire sur la situation de la voirie," Quebec (1941).

BERGERON, ARTHUR
"La Voirie dans la province de Québec," Mémoire pour la Commission royale d'enquête sur les problèmes constitutionnels, Québec (1953).

BIRD, W. R.
"History of the Highways of Nova Scotia," Public Archives of Nova Scotia.

BURTON, C. D.
"Memorandum on the Development of the Highway System Within the Province of Alberta," Edmonton (June 1964).

CAIRNS, H. L.
"Notes on the Road History of British Columbia," Department of Highways, British Columbia (1953).

CROSS, MICHAEL S.
"Some Aspects of Road Financing and Administration, 1791–1958," Department of Highways, Ontario.

"Crown Land Papers," Archives of Ontario.

DEAVILLE, STANLEY
"The Colonial Postal Systems of Vancouver Island and British Columbia, 1849–1871," Department of Highways, British Columbia.

DOUGLAS, SIR HOWARD
(Governor of New Brunswick)
"Letter to the Duke of Wellington, January 2, 1827, concerning the improvement of roads and military posts in Nova Scotia and New Brunswick to meet the threat of American expansion from Maine,"

Webster Collection, New Brunswick Museum, Saint John.

DUBE, YVES
"Les Problèmes administratifs et financiers de la voirie dans la province de Québec," Etude présentée par l'Union des Municipalités de la province de Québec à la Commission royale d'enquête sur les problèmes constitutionnels Québec (1955).

EVANS, R. D.
"Transportation and Communication in Nova Scotia, 1815–1850," M.A. thesis prepared at Dalhousie University under the direction of Professor D. C. Harvey, formerly Nova Scotia Provincial Archivist. See section II of this bibliography for an extension of this work.

"Experiences of Durand, mail courier, Halifax to Quebec," Public Archives, Ottawa.

FORTUNE, A. L.
"Overland to Cariboo," Department of Highways, British Columbia.

FOTHERGILL, CHARLES
"A Few Notes Made on a Journey from Montreal through the Province of Upper Canada in Feby 1817," Douglas Robertson collection, on loan to the Royal Ontario Museum. Part of this journal, which describes a carriole trip over the newly opened Kingston Road, is printed in Guillet, *Pioneer Inns and Taverns*, vol. 4, pp. 153–170.

GIFFORD, CHARLES
"A Few Remarks on a Tour in Upper Canada in 1837," in private possession, printed in Guillet, *The Pioneer Farmer and Backwoodsman*, vol. 2, pp. 344–358.

GILCHRIST, C. W.
"The Role of Roads in Canada's Development," Lecture, McGill University Extension Department (1953).

GILCHRIST, C. W.
"Social and Economic Aspects of Roads and Road Transport in Canada," paper presented to the International Road Federation World Meeting, Rome, 1955.

GUILLET, EDWIN C.
"Plank Roads," 1952, in larger libraries in Ottawa, Montreal, Toronto.

"Historical Materials on Road Construction in Manitoba," Department of Public Works, Winnipeg.

"Historical Notes on the Highways of Nova Scotia," Department of Highways, Nova Scotia.

"History of the Department of Roads," Department of Roads, Quebec.

"James Brown's Journal," Archives Department of the New Brunswick Museum, Saint John.

MACKENZIE, H. R.
"Highway Construction in Saskatchewan and Development and Influence of Highway Transportation," speech to the Ontario Good Roads Association, February 1929.

"Materials on the History of Alberta Roads," Department of Highways, Edmonton.

"Materials on the History of Roads in Saskatchewan," Department of Highways, Regina.

METHERAL, V. R.
"History of Highway Development in Saskatchewan during the Past 35 Years with Some of the Problems and Immediate Plans for the Future," Regina (May 1955).

MURPHY, DENIS
"The Building of the Cariboo Road," Department of Highways, British Columbia.

MURPHY, M. P., & CREWE, N. C.
"Memorandum on Some Early References to Newfoundland Roads," Archives of Newfoundland.

"Notes by E. C. Goulding re Trails of Manitoba (1864–1889)," Provincial Archives, Manitoba.

PATTISON, IRMA E.
"Historical Chronology of Highway Legislation in Ontario, 1774–1961," Department of Highways, Ontario (April 1964).

PETTY, THOMAS
"Trails of Alberta," Glenbow Foundation, Calgary.

PETTY, THOMAS
"Trails of Manitoba," Glenbow Foundation, Calgary.

PETTY, THOMAS
"Trials of Saskatchewan," Glenbow Foundation, Calgary.

"Public Works Documents," Public Archives, Ottawa.

"Report Q 167 re Roads," Public Archives, Ottawa.

"Road Report of November 1962," Department of Highways, Prince Edward Island.

ROBERTSON, W. G.
"History of the Trans-Canada Highway," address to the Canadian Good Roads Association in 1940.

DE ROTTENBURG, MAJOR BARON
"Map of the Principal Communications in Canada West," 1850–51, Public Archives, Ottawa.

SMITH, A. T.
"Transportation and Communication in Nova Scotia, 1749–1815," M.A. thesis prepared at Dalhousie University under the direction of Professor D. C. Harvey, formerly Nova Scotia Provincial Archivist.

VALLIS, DAVID A.
"A History of Canadian Highways," B.Sc. thesis, University of New Brunswick.

"La Voirie Provinciale," speech by Hon. Bernard Pinard, Ministre de la Voirie de la province de Québec, March 19, 1963.

II MAGAZINE AND NEWSPAPER ARTICLES, PAMPHLETS

ALDERMAN, TOM
"It's a Nuisance" (muskeg), *Imperial Oil Review* (June 1965).

BOWES, JIM
"The Canada Road," *Imperial Oil Review* (September 1952).

BOWES, JIM
"Everyone Lives by the Road," *Imperial Oil Review* (June 1961).

BOWES, JIM
"Mail Run from Mile Zero,"
Imperial Oil Review (April 1960).

CARON, I.
"Histoire de la voirie dans la
province de Québec," *Bulletin des
Recherches Historiques*, vol. 39.

COUSINS, G. V.
"Early Transportation in Canada,"
University Magazine, vol. 8.

CREECH, E. P.
"Brigade Trails of B.C.,"
The Beaver (March 1953).

CREECH, E. P.
"Similkameen Trails, 1846–61,"
*British Columbia Historical
Quarterly* (October 1941).

CRUIKSHANK, E. A.
"Early Traders and Trade Routes in
Ontario and the West, 1760–1783,"
*Transactions of the Canadian
Institute*, ser. 4, vols. 3 & 4.

DRAPER, W. N.
"Early Trails and Roads in the Lower
Fraser Valley," *British Columbia
Historical Quarterly* (January 1943).

DRAPER, W. N.
"Some Early Roads and Trails in
New Westminster District,"
British Columbia Historical Quarterly
(January 1945).

EVANS, R. D.
"Stage Coaches in Nova Scotia, 1815
to 1867," *Collections of the Nova
Scotia Historical Society*, vol. 24,
(1938), pp. 107–134.

FULTON, W. J.
"Highway High Jinks," *DHO News*,
Department of Highways, Ontario.

HARRINGTON, LYN
"Yukon's Wilderness Roads,"
Canadian Geographical Journal
(August 1962).

"Highway Destruction in Alaska,"
American Highways, January 1965.

"Highways and Highway
Construction," Feature Report,
Financial Post (March 27, 1965).

JEANNERET, M., & HARMAN, E. T.
"Frontier Trails," *Yours for the
Driving*, Ford Motor Co. (2nd
quarter, 1962).

JEANNERET, M., & HARMAN, E. T.
"Paddle Power," *Yours for the
Driving*, Ford Motor Co. (1st
quarter, 1962).

JEANNERET, M., & HARMAN, E. T.
"No Faster Than a Horse," *Yours
for the Driving*, Ford Motor Co.
(3rd quarter, 1962).

MACBRIDE, WILLIAM D.
"Yukon Stage Line," *The Beaver*
(June 1953).

MATHER, KEITH B.
"Why Do Roads Corrugate?"
Scientific American (January 1963).

METCALFE, BOB
"Goodbye to the Big Bend,"
Imperial Oil Review (August 1962).

MILLAR, H. M.
"Avalanche Control," *Canadian
Geographical Journal* (June 1960).

MILLMAN, THOMAS
"Diary of the Assistant Surgeon of
the B.N.A. Boundary Survey, 1873,"
*Transaction 26 of the Women's
Canadian Historical Society of
Toronto* (1927–28).

MOBERLY, WALTER
"History of the Cariboo Wagon
Road," *Art, Historical, and
Scientific Association*, Vancouver
(session 1907–08).

NUTE, GRACE LEE
"On the Dawson Road,"
The Beaver (Winter 1954).

OAKES, GARRETT
"Pioneer Sketches," *St. Thomas
Weekly Home Journal* (1876–77).

OLIVER, E. H.
"The Settlement of Saskatchewan to
1914," *Proceedings and Transactions
of the Royal Society of Canada*
(1956).

OSBORNE, A. C.
"The Migration of Voyageurs from
Drummond Island to
Penetanguishene," *Papers and
Records of the Ontario Historical
Society*, vol. 3.

PATTISON, IRMA
"Roads in Canada," *DHO News*,
Department of Highways, Ontario
(September 1963).

"Road Administration in Canada."
Canadian Good Roads Association,
Ottawa (1965).

Road and Wheel (January 1952)
This periodical bulletin of the
Canadian Good Roads Association
contains much material on historical
and contemporary events.

*Road Research Needs in Canada:
1965*, Technical Publication no. 27,
Canadian Good Roads Association,
Ottawa (1965).

ROBERTS, PHOEBE
"Diary of a Quaker Missionary
Journey to Upper Canada," ed. by
Leslie R. Gray, *Ontario History*,
vol. 42.

SMITH, G. I.
"Good Roads in the Atlantic
Provinces," *Atlantic Advocate*
(April 1961).

SPRAGGE, GEORGE
"Colonization Roads in Canada
West, 1840–1867," *Ontario History*,
vol. 49.

STANLEY, G. D.
"Medical Pioneering in Alberta,"
The Early West (1957).

STEAD, R. J. C.
"Highways of British Columbia,"
Canadian Geographical Journal
(August 1947).

*This is Our Golden Anniversary,
1914–1964*, Canadian Good Roads
Association, Ottawa (1964).

THOMPSON, JOHN
"Travelling the Main Road of
Yesteryear," *Lake of Two Mountains
Gazette* (July 30, 1964).

VAN STEEN, MARCUS
"The Trans-Canada Highway Makes
Road-Building History North of
Lake Superior," *Canadian
Geographical Journal* (November
1962).

"Walter Moberly's Report on the
Roads of British Columbia, 1863,"
*British Columbia Historical
Quarterly* (January 1945).

WESLEY, GORDON
"Romkey & Co." *Imperial Oil
Review* (February 1965).

WOLSELEY, VISCOUNT
"Narrative of the Red River
Expedition," *Blackwood's
Magazine* (1870–71).

III BOOKS

ALEXANDER, SIR JAMES
*L'Acadie, or Seven Years'
Explorations in British America*,
2 vols., London, 1849.

ANONYMOUS
Journal of a Trip through Canada,
Ottawa, 1931.

D'ARTIGUE, JEAN
*Six Years in the Canadian
North-West*, Toronto, 1882.

"A. S."
A Summer Trip to Canada,
London, 1846.

BEAVEN, JAMES
Recreations of a Long Vacation,
London, 1846.

BEGG, ALEXANDER
*Seventeen Years in the Canadian
North-West*, London, 1884.

BÉRARD, MICHEL
Les Routes du Québec, Quebec,
1964.

*BERTON, PIERRE
*The Klondike: The Life and Death
of the Last Great Gold Rush*,
Toronto, 1958.

BIGSBY, J. J.
*The Shoe and Canoe, or Pictures of
Travel in the Canadas*, 2 vols.,
London, 1850.

BONNYCASTLE, SIR RICHARD
Canada and the Canadians, 2 vols.,
London, 1846.

BOUCHETTE, JOSEPH
*The British Dominions in North
America*, 2 vols., London, 1831.

BUCKINGHAM, JAMES
*Canada, Nova Scotia, New
Brunswick*, London, 1843.

BUTLER, SIR W. F.
The Great Lone Land, London, 1924.

Canadian Good Roads Association
*Geometric Design Standards for
Canadian Roads and Streets*,
Ottawa, 1963.

Canadian Tax Foundation
Taxes and Traffic, Toronto, 1955.

CANNIFF, WILLIAM
*History of the Settlement of Upper
Canada*, Toronto, 1869.

CHAMPLAIN, SAMUEL DE
Voyages, ed. E. F. Slafter, the
Prince Society, 2 vols., Boston,
1878–82.

CHURCH, H. E.
*An Emigrant in the Canadian
Northwest*, London, 1929.

COKE, E. T.
A Subaltern's Furlough, 2 vols.,
New York, 1833.

COLMER, J. G.
Across the Canadian Prairies,
London, 1895.

CUMMINGS, H. R.
Early Days in Haliburton,
Toronto, 1963.

DARLING, W. STEWART
Sketches of Canadian Life,
London, 1849.

DE ROS, F. F.
*Personal Narrative of Travels in
1826*, London, 1827.

DICKENS, CHARLES
American Notes, London, 1842.

*DOUGLAS, GEORGE M.
Lands Forlorn, New York, 1914.

DOWNS, ART
*Wagon Road North: Historic
Photos of the Cariboo Gold Rush*,
Quesnel, B.C., 1960.

DUNLOP, WILLIAM
Recollections of the American War,
Toronto, 1905.

DUNLOP, WILLIAM
Two and Twenty Years Ago,
Toronto, 1859.

ECCLES, W. J.
*Canada Under Louis XIV,
1663–1701*, Toronto, 1964.

*Eighty Years' Progress in British
North America*, Toronto, 1863.

ELKINGTON, W. M.
Five Years in Canada, London,
1895.

ENGLAND, ROBERT
*The Colonization of Western
Canada*, London, 1936.

FERGUSSON, ADAM
*Practical Notes Made during a
Tour*, Edinburgh, 1834.

The First Fifty Years, 1914–1964
(of the American Association of
State Highway Officials), 1965.

FIRTH, EDITH
The Town of York, 1793–1815,
Toronto, 1962.

FLANDRAU, GRACE
Red River Trails, St. Paul (?),
19 ? .

FONSECA, W. G.
On the St. Paul Trail in the Sixties,
Winnipeg, 1900.

FOWLER, THOMAS
*Journal of a Tour through British
America*, Aberdeen, 1832.

FRASER, SIMON
*Letters and Journals of Simon
Fraser, 1806–1808*, ed. W. Kaye
Lamb, Toronto, 1960.

*Functional Report: Proposed
Reconstruction of Highway 401 in
Metropolitan Toronto*, Toronto,
1963.

GEIKIE, JOHN C.
*George Stanley, or Life in the
Woods*, London, 1864.

*GLAZEBROOK, G. P.
*A History of Transporation in
Canada*, Toronto, 1938; reissued as
no. 11 and 12, Carleton Library,
Toronto, 1964.

GODSELL, PHILIP
*The Romance of the Alaska
Highway*, Toronto, 1944.

GOLDIE, JOHN
*Diary of a Journey through Upper
Canada*, Toronto, 1897.

GOURLAY, ROBERT
*A Statistical Account of Upper
Canada*, 3 vols., London, 1822.

GREEN, ANSON
Life and Times, Toronto, 1877.

*GUILLET, EDWIN C.
Early Life in Upper Canada,
Toronto, 1933.

*GUILLET, EDWIN C.
The Great Migration, New York
& Toronto, 1937.

*GUILLET, EDWIN C.
Pioneer Inns and Taverns, 5 vols.,
Toronto, 1954–1962.

*GUILLET, EDWIN C.
*Toronto: From Trading-Post to
Great City*, Toronto, 1934.

HAIGHT, CANNIFF
*Country Life in Canada Fifty
Years Ago*, Toronto, 1885.

HALL, BASIL
*Travels in America in 1827 and
1828*, 3 vols., Edinburgh, 1829.

HALL, MRS. CECIL
*A Lady's Life on a Farm in
Manitoba*, London, 1884.

HALL, FRANCIS
Travels in Canada in 1816–17,
London, 1818.

HAMILTON, J. C.
The Prairie Province, Toronto, 1876.

HARMAN, BRUCE
*'Twas 26 Years Ago: Narrative of
the Red River Expedition*,
Toronto, 1896.

HATTON, J.
Today in America, 1881.

HEAD, SIR FRANCIS B.
The Emigrant, London, 1846.

HEAD, SIR GEORGE
*Forests Scenes and Incidents in the
Wilds of North America*,
London, 1829.

*HEARNE, SAMUEL
*A Journal from Hudson's Bay to
the Northern Ocean in the Years
1769–1772*, London, 1795.

*HENRY, ALEXANDER
*Travels and Adventures in Canada,
1760–76*, New York, 1809.

HERIOT, GEORGE
Travels through the Canadas,
London, 1808.

HIEMSTRA, MARY
Gully Farm, Toronto, 1955.

HILL, A. S.
*From Home to Home: Autumnal
Wanderings in the North-West*,
London, 1885.

HIND, H. Y.
British North America, Toronto,
1851.

HOLLOWAY, R. E.
*Through Newfoundland with the
Camera*, St. John's, 1905.

HOOPLE, CARRIE MUNSON
*Along the Way with Pen and
Pencil*, New York, 1909.

HOWISON JOHN
Sketches of Upper Canada,
Edinburgh, 1821.

HUYSHE, G. L.
The Red River Expedition,
London, 1871.

L'Ile D'Orléans, Historic
Monuments Commission of Quebec,
Quebec, 1928.

*INNIS, H. A. (ED.)
The Dairy Industry in Canada,
Toronto, 1937.

*INNIS, H. A. (ED.)
The Fur Trade in Canada,
Toronto, 1927.

*INNIS, H. A. (ED.)
*Select Documents in Canadian
Economic History*, 2 vols.,
Toronto, 1929–33.

INNIS, HAROLD A., HARVEY, D. C.,
& FERGUSSON, CHARLES B.
*The Diary of Simeon Perkins
(1766–1796)*, Champlain Society,
1948, 1958, 1961.

INNIS, MARY QUAYLE (ED.)
The Diary of Mrs. Simcoe,
Toronto, 1965.

JAMESON, ANNA
*Winter Studies and Summer Rambles
in Canada*, 3 vols., London, 1838.

JOHNSON, R. B.
Very Far West Indeed, London,
1872.

JOHNSTON, J. F. W.
Notes on North America,
Edinburgh, 1851.

JOHNSTONE, C. L.
*Winter and Summer Excursions in
Canada*, London, 1894.

*KANE, PAUL
*Wanderings of an Artist among the
Indians of North America*,
London, 1859.

KAVANAGH, MARTIN
The Assiniboine Basin,
Winnipeg, 1946.

[KING, MRS. H. B.]
*Letters from Muskoka by an
Emigrant Lady*, London, 1878.

KINGSFORD, WILLIAM
*History, Structure, and Statistics of
Plank Roads*, Philadelphia, 1852.

KINGSTON, W. H. G.
*Western Wanderings, or a Pleasure
Tour in the Canadas*, London, 1856.

LANGLEY, A. J.
*A Glance at British Columbia and
Vancouver's Island in 1861*,
London, 1862.

LANGTON, JOHN
Early Days in Upper Canada,
Toronto, 1926.

LAROCHEFOUCAULD-LIANCOURT,
DUC DE
Travels, 2 vols., London, 1799.

LAUT, AGNES
The Cariboo Trail, Toronto, 1916.

LESSARD, J. C.
Transportation in Canada, Royal
Commission on Canada's Economic
Projects, 1956.

LIZARS, K. M.
The Valley of the Humber,
Toronto, 1913.

LOWER, A. R. M.
*Canadians in the Making: A Social
History of Canada*, Toronto, 1958.

*LOWER, A. R. M.
*Settlement and the Forest Frontier
in Eastern Canada*, Toronto, 1936.

LYELL, SIR CHARLES
*Travels in North America in the
Years 1841–2*, 2 vols., New York,
1845.

*McCOURT, EDWARD
The Road across Canada,
Toronto, 1965.

MACDERMOTT, C. L.
*Facts for Emigrants, a Journey
from London*, London, 1868.

McDONALD, ARCHIBALD
*Peace River, A Canoe Voyage
from Hudson's Bay to the Pacific*,
Ottawa, 1872.

M'DONALD, JOHN
Narrative of a Voyage,
London, 1822.

MACDONELL, ALLAN
The North-West Transportation,
Toronto, 1958.

McDOUGALL, JOHN
Forest, Lake, and Prairie,
Toronto, 1895.

McDOUGALL, JOHN
On Western Trails, Toronto, 1911.

McDOUGALL, JOHN
Pathfinding on Plain and Prairie,
Toronto, 1897.

McDOUGALL, JOHN
Saddle, Sled, and Snowshoe,
Toronto, 1896.

MacEWAN, GRANT
Between the Red and the Rockies,
Toronto, 1952.

MacGOWAN, MICHAEL
The Hard Road to Klondike,
London, 1962.

McKAY, R. W. S.
*Travellers' Guide to the St.
Lawrence,* Montreal, 1845.

MACKENZIE, SIR ALEXANDER
General History of the Fur Trade,
London, 1801.

*MACKENZIE, SIR ALEXANDER
*Voyage from Montreal through
North America,* London, 1801.

McNAUGHTON, MARGARET
Overland to Cariboo in 1862,
Toronto, 1896.

MACTAGGART, JOHN
Three Years in Canada, London,
1829.

MacNUTT, W. S.
New Brunswick, Toronto, 1963.

MASSON, L. R.
*Les Bourgeois de la Compagnie du
Nord-Ouest,* Quebec, 1889–90.

MILTON, VISCOUNT, &
CHEADLE, W. B.
The North-West Passage by Land,
London, 1865.

MOODIE, SUSANNA
Roughing it in the Bush,
London, 1852.

MORTON, A. S.
*A History of the Canadian West
to 1870–1,* London, 1939.

MORTON, W. L.
Manitoba: A History, Toronto, 1957.

MURRAY, SIR C. A.
Travels in North America,
London, 1839.

MURRAY, FLORENCE
*Muskoka and Haliburton
1615–1875,* Toronto, 1963.

[NEED, THOMAS]
Six Years in the Bush,
London, 1838.

NIX, JAMES E.
Mission among the Buffalo,
Toronto, 1960.

PALLISER, JOHN
*The Journals: Detailed Reports by
Captain Palliser,* London, 1859–63.

*PALLISER, JOHN
*Solitary Rambles and Adventures
of a Hunter in the Prairies,*
London, 1853.

PEEL, BRUCE
*A Bibliography of the Prairie
Provinces to 1953,* Toronto, 1956.

PEEL, BRUCE
*Supplement to A Bibliography of
the Prairie Provinces to 1953,*
Toronto, 1963.

PERLIN, A. B.
*The Story of Newfoundland,
1497–1959,* St. John's, 1959.

PICKEN, ANDREW
The Canadas, London, 1832.

PICKERING, JOSEPH
Inquiries of an Emigrant,
London, 1831.

[PORTER, JANE]
*A Six Weeks' Tour in Western
Canada,* Montreal, 1865.

PRINGLE, J. F.
*Lunenburgh, or the Old Eastern
District,* Cornwall, 1890.

PROWSE, D. W.
A History of Newfoundland,
London, 1895.

PRYOR, ABRAHAM
*An Interesting Description of
British America from Personal
Knowledge and Observation,*
Providence, 1819.

RAE, W. F.
Newfoundland to Manitoba,
New York, 1881.

*Report of the Proceedings of the
First Annual Convention of the
Manitoba Good Roads Association,*
Winnipeg, 1910.

"A RETURNED DIGGER"
*Cariboo, the Newly Discovered
Gold Fields of British Columbia,*
London, 1862.

RICHMAN, THOMAS
*Notes of a Short Visit to Canada
and the States,* London, 1886.

ROBERTSON, DOUGLAS (ED.)
An Englishman in America, 1785,
Toronto, 1933.

ROBERTSON, JOHN ROSS (ED.)
*The Diary of Mrs. John Graves
Simcoe,* Toronto, 1911.

*ROBINSON, PERCY
Toronto during the French Régime,
Toronto, 1933.

ROE, F. G.
The North American Buffalo,
Toronto, 1951.

ROPER, EDWARD
*By Track and Trail through
Canada,* London, 1891.

ROWAN, J. J.
Emigrant and Sportsman in Canada,
London, 1876.

ROWE, F. W.
*The Development of Education in
Newfoundland,* Toronto, 1964.

ROY, THOMAS
*Remarks on the Principles of
Roadmaking as Applicable to
Canada,* Toronto, 1841.

*Royal Bank of Canada Monthly
Letters* (current).

RUSSELL, R. C.
The Carlton Trail, Saskatoon, 1955.

SCHOOLCRAFT, HENRY
The Indian Tribes, 6 vols.,
Rochester, 1851–6.

SCHULTZ, JOHN
The Old Crow Wing Trail,
Winnipeg, 1894.

[SHERK, M. G.]
*Pen Pictures of Early
Pioneer Life in Upper Canada,*
Toronto, 1905.

SHIRREFF, PATRICK
A Tour through North America,
Edinburgh, 1835.

SHORTT, ADAM, & DOUGHTY, A. G.
(ED.)
Canada and Its Provinces, 23 vols.,
Edinburgh & Toronto, 1914–17.

SIMPSON, SIR GEORGE
*Fur Trade and Empire; George
Simpson's Journal,* London, 1931.

SLEIGH, B. W. A.
*Pine Forests and Hacmatack
Clearings,*
London, 1853.

SMALLWOOD, J. R. (ED.)
The Book of Newfoundland, 2 vols.,
St. John's, 1937.

SMITH, D. W.
*A Short Topographical Description
of Upper Canada,* London, 1799.

SMITH, WILLIAM
*History of the Post Office in
British North America,*
Cambridge, Mass., 1920.

*Southern Manitoba and Turtle
Mountain Country,* 1880.

SQUIRES, W. A.
*The 104th Regiment of Foot,
1803–17,* Fredericton, 1962.

STANSBURY, PHILIP
*A Pedestrian Tour of Two
Thousand Three Hundred Miles in
North America,* New York, 1822.

STEPHENS, C. A.
*The Adventures of Six Young Men
in the Wilds of Maine and Canada,*
London, 1885.

STRICKLAND, SAMUEL
*Twenty-seven Years in Canada
West,* 2 vols., London, 1853.

STUART, CHARLES
*The Emigrant's Guide to Upper
Canada,* London, 1820.

"A SUCCESSFUL DIGGER"
*The Wonders of the Gold-Diggings
of British Columbia,* London, 189?.

TRAILL, CATHARINE P.
The Backwoods of Canada,
London, 1836.

TUDOR, HENRY
*Narrative of a Tour in North
America, 1831–2,* 2 vols.,
London, 1834.

*TYRRELL, JAMES W.
*Across the Sub-Arctics of Canada,
a Journey of 3200 miles by Canoe
and Snowshoe through the Barren
Lands,* Toronto, 1897.

*WADE, M. S.
The Overlanders of '62 (Memoir IX,
Archives of British Columbia),
Victoria, 1931.

WASHBURN, STANLEY
*Trails, Trappers, and Tenderfeet
in the New Empire of Western
Canada,* London, 1912.

WATKIN, SIR E. W.
*Canada and the States:
Recollections,* London, 1887.

WELD, ISAAC
*Travels through the States of
North America,* London, 1799.

*WILBY, THOMAS W.
A Motor Tour through Canada,
London, 1914.

WILLIAMS, G. B.
The Trans-Canada Highway, 1957.

WILLIS, N. P.
Canadian Scenery, 2 vols.,
London, 1842.

WINCH, DAVID M.
*The Economics of Highway
Planning,* Toronto, 1963.

YOUNG, E. R.
*By Canoe and Dog Train among
the Cree and Salteaux Indians,*
London, 1890.

MAPS on pages 9, 27, 36, 41, 96, 104, 213, and 230 were prepared by Robert Kunz. Diagrams on pages 14, 64, 66, 68, 156, 202, and 203 were prepared by A. D. Margison and Associates Limited. Pictures of construction equipment in chapter 14 were obtained through the courtesy of the Canadian Association of Equipment Distributors. In the following paragraph these abbreviations are used: CGRA, Canadian Good Roads Association; CRTA, Canadian Road Transport Archives; DHO, Department of Highways, Ontario; NFB, National Film Board; PA, Public Archives of Canada; ROM, Royal Ontario Museum, University of Toronto; TPL, Toronto Public Libraries. Illustrations on each page are cited from top to bottom, and from left to right.

i: ROM. ii: PA; Author's collection. iii: Saskatchewan Government photograph. iv: Department of Public Works, New Brunswick; TPL, John Ross Robertson Collection. v: Saskatchewan Archives. vi: CRTA. vii: Author's collection. viii: Canadian Clark Limited. ix: Toronto Transit Commission. x: Office Provincial de Publicité, Québec. 1: Bruno Engler, Alpine Films; National Film Board. 2: National Gallery of Canada. 4: ROM. 5: PA. 7: B. M. Litteljohn collection. 8: PA. 9: PA. 10: ROM. 11: N. P. Willis, ed., *Canadian Scenery* (1842). 12: National Gallery of Canada. 14: Office Provinciale de Publicité, Québec, Musée de la Province. 15: National Gallery of Canada. 16: Author's collection. 17: ROM. 18: ROM; PA. 19: PA. 20: ROM; PA. 23: PA. 24: ROM. 26: R. E. Holloway, *Through Newfoundland with the Camera* (1905). 29: ROM. 34: New Brunswick Museum, Saint John; ROM. 35: New Brunswick Museum, Saint John. 36: ROM. 38: PA. 40: Egerton Ryerson, *The Story of My Life* (1883). 43: PA; PA. 45: ROM. 47: Tremaine Map of the County of Peel. 48: P.A. 49: TPL, John Ross Robertson Collection. 50: TPL. 52: G. H. Needler collection. 53: Bank of Montreal, Picton, Ontario. 55: ROM. 56: Bruce Harman, *'Twas Twenty-six Years Ago* (1896). 58: TPL, John Ross Robertson Collection; TPL. 60: W. C. Bryant, *Picturesque America* (1872). 61: TPL. 62: Ontario Archives; *Canadian Illustrated News* (1871). 67: *Eighty Years' Progress in British North America* (1863). 70: N. P. Willis, ed., *Canadian Scenery* (1842). 72: Ontario Archives; TPL; Province de Québec, Musée de la Province. 73: PA; TPL, John Ross Robertson Collection; PA. 74: PA. 76: G. M. Grant, ed., *Picturesque Canada* (1880). 77: G. M. Grant, ed., *Picturesque Canada* (1880); G. M. Grant, ed., *Picturesque Canada* (1880). 78: N. P. Willis, ed., *Canadian Scenery* (1842). 79: Ralph Greenhill collection; TPL, John Ross Robertson Collection. 81: Perkins Bull collection. 82: ROM. 84: Author's collection; Ontario Archives; *Harper's Weekly* (1870). 89: TPL. 90: The Bettman Archive. 92: PA. 94: British Columbia Provincial Archives; British Columbia Provincial Archives. 95: British Columbia Provincial Archives. 96: British Columbia Provincial Archives. 98: Vancouver City Archives; British Columbia Provincial Archives. 99: Vancouver City Archives. 100: Edward Roper, *By Track and Trail through Canada* (1891). 102: PA. 104: Hudson's Bay Company. 105: Geological Survey of Canada. 107: PA. 108, 109: cartoons from *Canadian Illustrated News* (1871); painting on 109: ROM. 110: PA. 111: G. M. Grant, ed., *Picturesque Canada* (1880). 112: Department of Mines and Technical Surveys. 115: The Bettman Archive. 116: T. W. Wilby, *A Motor Tour through Canada* (1914). 118: ROM. 120: G. M. Grant, ed., *Picturesque Canada* (1880). 121: Ralph Greenhill collection; PA. 123: Toronto Transit Commission; Saskatchewan Archives; *Porcupine [Ontario] Gold Rush Souvenir Booklet*. 124: George Eastman House. 125: Geological Survey of Canada; Ralph Greenhill collection. 126: National Gallery of Canada; PA; ROM; CRTA. 127: *Canadian Illustrated News* (1881); *Canadian Illustrated News* (1871). 128: Dr. W. C. Givens collection; Ralph Greenhill collection; CRTA. 129: Ontario Hydro; DHO. 130: C. P. Hornung, *Wheels across America*, A. S. Barnes (1959). 132: Ralph Greenhill collection. 133: R. E. Holloway, *Through Newfoundland with the Camera* (1905); Florence Murray collection. 134: PA. 135: Russell Foster collection. 136: A. A. Silcox collection. 137: PA. 138: Ontario Motor League; CRTA. 139: TPL, John Ross Robertson Collection. 140: Department of Highways, Nova Scotia; W. S. Herrington, *Lennox and Addington* (1913); Ralph Greenhill collection; British Columbia Provincial Archives; PA, Sir Sandford Fleming Collection; *Canadian Geographical Journal* (1932). 141: Department of Public Works, New Brunswick; Province de Québec; Department of Public Works, New Brunswick. 142: Automotive History Collection, Detroit Public Library. 144: U. H. Dandurand collection; G. M. Grant, ed., *Picturesque Canada* (1880). 145: *Scientific American* (1869); Hamilton *Spectator*. 146: CRTA. 147: Ontario Motor League; CRTA. 148: PA. 149: CRTA; *Canadian Motorist* (1930), Ontario Motor League. 150: Automotive History Collection, Detroit Public Library. 151: Automotive History Collection, Detroit Public Library. 152: U. H. Dandurand collection. 153: U. H. Dandurand collection. 154: James Collection of Early Canadiana. 157: CRTA. 158: Ontario Motor League. 159: Annual Report on Highway Improvement, Ontario, 1920. 160: Ontario Motor League. 165: Author's collection; Canada Cement Company. 167: CRTA. 169: Manitoba Archives. 171: James Montagnes collection; James Montagnes collection; DHO. 172: Saskatchewan Archives; CGRA. 173: Saskatchewan Archives;

Annual Report on Highway Improvement, Ontario, 1909; Saskatchewan Archives. 174: Department of Public Works, New Brunswick; PA. 175: S. T. Franks collection; Department of Public Works, New Brunswick. 176: Annual Report on Highway Improvement, Ontario, 1912; PA; CGRA. 177: CGRA; CGRA; Department of Public Works, Manitoba; DHO. 178: NFB. 180: Ernest Brown Collection, Department of Industry and Development, Alberta; William D. MacBride collection. 181: William D. MacBride collection; William D. MacBride collection. 182: Author's collection. 183: NFB. 184: NFB. 185: NFB. 186: NFB. 188: NFB. 189: NFB. 191: NFB. 192: DHO. 194: DHO; DHO. 195: DHO. 197: Nova Scotia Information Service. 198: Department of Highways, British Columbia; NFB; Iowa Manufacturing Company; Koehring Waterous Limited; Bucyrus-Erie Company of Canada. 199: LW Manufacturing Company; Department of Public Works, Manitoba. 200: Canadian Clark Limited. 201: NFB. 204: NFB. 205: NFB. 206: DHO. 207: Overland Transport Co.; DHO. 209: Ontario Provincial Police. 210: Office Provincial de Publicité, Québec. 211: Ontario Provincial Police. 214: ROM; DHO; Office Provincial de Publicité, Québec. 215: Nova Scotia Information Service. 216: DHO; Department of Highways, Newfoundland; Saskatchewan Archives; NFB. 217: British Columbia Government photograph; NFB. 218: Canadian Government Travel Bureau. 220: T. W. Wilby, *A Motor Tour through Canada* (1914). 222: T. W. Wilby, *A Motor Tour through Canada* (1914). 224: CRTA. 225: CRTA; DHO. 226: DHO; Saskatchewan Photo Service. 227: James Montagnes collection. 228: CGRA; CGRA; CGRA; Bruno Engler, CGRA. 323: Nova Scotia Information Service. 233: Department of Highways, British Columbia. 246: DHO.

Listings refer both to the text and to the illustration captions.

Index

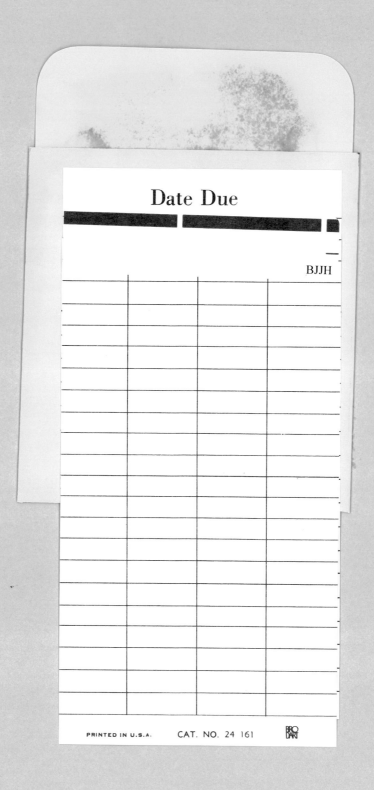

Date Due

BJJH

PRINTED IN U.S.A.   CAT. NO. 24 161   BRO DART